Workbook to Accompany

Statistics: Concepts and Applications for Science

David C. LeBlanc

Ball State University

JONES AND BARTLETT PUBLISHERS

Sudbury, Massachusetts

BOSTON   TORONTO   LONDON   SINGAPORE

World Headquarters
Jones and Bartlett Publishers
40 Tall Pine Drive
Sudbury, MA 01776
978-443-5000
info@jbpub.com
www.jbpub.com

Jones and Bartlett Publishers Canada
2406 Nikanna Road
Mississauga, ON L5C 2W6
CANADA

Jones and Bartlett Publishers International
Barb House, Barb Mews
London W6 7PA
UK

Cover image © Sandy Felsenthal / Corbis

ISBN 0-7637-2220-0

Executive Editor: Stephen Weaver
Managing Editor: Dean DeChambeau
Production Manager: Amy Rose
Marketing Manager: Matthew Bennett
Associate Editor: Rebecca Seastrong
Production Assistant: Tracey Chapman
Production Coordination: Jennifer Bagdigian
Manufacturing and Inventory Control: Therese Bräuer
Cover Design: Night & Day Design
Text Design: Anne's Books
Composition: Northeast Compositors
Technical Art: George Nichols
Printing and Binding: Courier Stoughton
Cover Printing: Courier Stoughton

.

Printed in the United States of America
08 07 06 05 04   10 9 8 7 6 5 4 3 2 1

Contents

What Is "Data" and Where Does It Come From?

Homework Problems

1. Create a "data table" that lists subjects, variables, and values for a study of the relationship between *body structure* and *blood cholesterol* in men from the descriptive information given below. Give each variable an appropriate name. It is possible that some information in the description below would not be appropriate for this data table.

   Joe is a 45-yr-old divorced father of three with brown hair and brown eyes. He is 183 cm tall and weighs 86 kg. His blood LDL (low density lipid, "bad") cholesterol is 150 mg/dl, and his HDL (high density lipid, "good") cholesterol is 45 mg/dl. His ratio of total to HDL cholesterol is 4.1. Scot is a 19-yr-old single college student studying to become a doctor. He has blonde hair and blue eyes and is a marathon runner. He is 190 cm tall and weighs 80 kg. His LDL cholesterol is 90 mg/dl, and his HDL cholesterol is 55 mg/dl. His ratio of total to HDL cholesterol is 2.9. Rocky is a 30-yr-old professional wrestler. He is 193 cm tall and weighs 115 kg. His LDL cholesterol is 195 mg/dl, and his HDL cholesterol is 75 mg/dl. His ratio of total to HDL cholesterol is 3.8.

2. For each of the following variables, indicate if it is quantitative or categorical. If it is quantitative, also indicate if it is discrete or continuous.

   a. Number of bacteria colonies on a petri plate _____

   b. Petal color of a flower _____

   c. Concentration of glucose in a blood sample _____

3. In the following paragraph, state whether each of the numbers in boldface is a *population parameter* or a *sample statistic*.

An investigator wants to determine if state universities have student bodies that more closely match the gender and ethnic composition of the U.S. population than those of private universities. He obtains gender and ethnicity data from a random sample of 100 state universities and 100 private universities. The average percentage of African-American students at the state universities was **8%**, while for the private universities the average was **4%**. The percentage of female students was **53%** and **55%** at state and private universities, respectively. According to the most recent U.S. census, **13%** of the U.S. population described themselves as African-American, and **51%** were women.

8%: _____   4%: _____   53%: _____

55%: _____   13%: _____   51%: _____

4. Answer the questions below based on histograms (A) to (D) of 10 measurements made on a standard sample with a known value of 10.

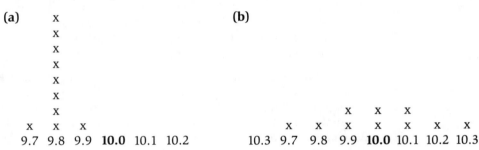

(a)
```
      x
      x
      x
      x
      x
      x
      x
  x   x   x
 9.7 9.8 9.9 10.0 10.1 10.2
```

(b)
```
                        x   x   x
                    x   x   x   x   x   x   x
                   10.3 9.7 9.8 9.9 10.0 10.1 10.2 10.3
```

(c)
```
              x
          x   x   x
          x   x   x
          x   x   x
     9.7  9.8  9.9 10.0 10.1 10.2 10.3
```

(d) x
```
    x   x
    x   x   x   x   x   x   x
   9.7 9.8 9.9 10.0 10.1 10.2 10.3
```

a. In which histograms (may be more than one) do the data values indicate *bias*? Explain.

b. In which one histogram do the data values indicate the highest *precision*? Explain.

c. In which one histogram do the data values indicate that the most accurate estimate of the true value (10) would be obtained? Explain.

5. Four hospitals in a major urban area do a coordinated screening program to determine the prevalence of high blood pressure in adults over 30 yrs old. The percentages of individuals screened at these four hospitals who had high blood pressure were 18, 25, 10, and 13%. Let's assume that the data obtained at all four hospitals were obtained using the same valid measurement protocol. Explain how the four samples of individuals measured at these hospitals could produce four different estimates of the one true prevalence of high blood pressure for the population in this urban area.

6. **a.** How could measurement error contribute to sampling variability, but not cause bias.

   b. Describe a situation in which measurement error results in bias.

7. Beginning on Line 101, Column A in the random numbers table (Appendix Table 1), reading down the column, and using the first three digits of each four-digit number, write the first 10 random numbers that fall in the range 0 to 500. Retain leading zeros to allow for random numbers between 0 and 99.

_____ _____ _____ _____ _____ _____ _____ _____ _____ _____

8. Describe how you would use a random numbers table to choose a simple random sample of 10 students from a large class.

9. **a.** What does it mean when we say a sample is *completely random*?

   b. Why is randomization critical to scientific sampling study design?

   c. If randomization is so critical to scientific sampling study design, why do many studies use systematic sampling rather than random sampling?

10. **a.** Why is replication critical to sampling study design?

    b. Given the importance of replication, what is the primary reason that many studies have inadequate replication?

11. Under what conditions should you consider using stratified random sampling? Explain.

12. What is the main purpose of randomization and replication in *experimental design,* as distinct from sampling design?

13. What aspect(s) of experimental design is (are) supposed to eliminate or account for the influence of nontreatment factors on the response of subjects? Explain.

14. You want to determine if bright, full-spectrum lights improve people's mood during the dark days of winter. Describe how you would use a random numbers table to randomly assign 20 of 40 volunteers to the treatment group (receives bright lights) and 20 to a control group (receives normal fluorescent lights).

15. A medical team wants to determine which form of mastectomy is most effective for prolonging life for women with breast cancer: radical mastectomy (removal of breast, chest muscles, and lymph nodes), simple mastectomy (removal of the breast only), or "lumpectomy" (removal of only the cancerous lump, leaving the breast tissue otherwise intact) plus chemotherapy? The team examines the records from five large hospitals and compares the survival times after surgery of all women who have had these various types of surgery for breast cancer.

   a. Was this study a true experiment? Explain your answer.

   b. What was the explanatory (treatment) variable and response variable in this study?

      Explanatory variable: _____

      Response variable: _____

   c. Identify a confounding factor that might obscure the connection between the treatment (explanatory) variable and the response variable. Describe how this extraneous variable might make it difficult to assess the relative effectiveness of the treatments.

16. Investigators in a human performance laboratory wanted to determine if multiple short periods of exercise per day provide the same benefit as a single long period of exercise. They obtain a sample of 36 overweight female college students. They randomly assign 12 of these women to each of three experimental groups. One group does a standard exercise regime 3 times per day, for 10 minutes each time. The second group exercises 2 times per day, for 15 minutes each time. The third group exercises once per day for 30 minutes. After three months, each woman is measured for how much weight they lost (in kilograms).

   a. Identify the experimental units, the explanatory (treatment) variable, and the response variable in this experiment.

      Experimental units: _____

      Explanatory variable: _____

      Response variable: _____

b. Draw a diagram that outlines this completely randomized study design. Indicate where randomization occurs, the size of the treatment groups, the nature of the treatments, and the response variable. Use the experimental design diagrams in the text as examples, but replace general terms with terms specific to this study.

c. Anytime you read the results of an experiment you should identify the larger population to which the results can be validly extrapolated. One way to make this identification is to ask the question, "From what population do the study subjects represent a representative sample?"

(1) Using the most stringent assessment, describe the larger population to which the results of this study can be generalized.

(2) Suppose you wanted to apply the results most broadly and were willing to accept a less stringent standard. Describe the larger population to which the results of the study could be generalized.

(3) Describe groups within the U.S. population to whom the results of this study should *not* be applied.

17. a. Describe in your own words the concept of *measurement validity*.

In each of the following cases (b)–(d), comment on the issue of measurement validity, based on issues of validity, accuracy, and precision.

b. pH paper was used to measure pH prior to the development of pH meters. The paper would change color in response to different pH's and was able to measure pH with precision of $\pm$ 0.1 pH unit. pH meters can attain precision of $\pm$ 0.001 pH units. Is using pH paper to measure pH of a solution "valid"? Is it more or less valid than using a pH meter? Explain.

c. You want to determine if the concentration of a salt solution is greater than or less than the concentration of a sugar solution but lack an analytical chemistry lab. Is it valid to use taste intensity (compare relative "saltiness" vs. "sweetness")? Explain.

d. Is the amount of a drug substance consumed by a person a valid measure of the "dose" in experiments to test the effect of the drug on some metabolic process?

18. Many controlled experiments to test the effectiveness of new drugs to treat cancer in humans use lab rats as study subjects. Although there are good reasons for using rats, such experiments lack realism. What is the meaning of the word "realism" in this context, and why do cancer drug experiments with rats lack realism?

19. What is the primary distinction between controlled experiments and natural experiments?

For each of the studies in Questions 20 and 21, describe the variable of interest, the population of interest, and the replicate units that make up this population. Describe if the original data values produced by the study design include pseudo-replication or lack of independence. Explain how the data might be manipulated to eliminate pseudo-replication (reduced to one data value per replicate) *or* how nonindependent measurements are related and how this problem could be fixed.

20. *Identifying Pseudo-Replication*

   a. *Study Question:* Does agricultural fertilizer runoff increase algae in freshwater lakes in Indiana? *Study Design:* Two similar size lakes are identified. One receives drainage water from agricultural fields; the other is in the middle of a wilderness area. Once each month during the period April through October, 10 five-liter water samples are taken at random locations on each lake. Algae biomass (grams dry weight/ml) is measured on three one-liter subsamples taken from each five-liter water sample.

   Variable of Interest: _____

   Population of Interest: _____

   Replicate: _____

   Pseudo-replication? Yes / No   Lack of Independence?   Yes / No

   Explanation: _____

   _____

   _____

   b. *Study Question:* What is the effect of fertilizer inputs on algae production in artificial freshwater ponds? *Study Design:* Twenty ponds of equal surface area and depth are dug at a facility near Purdue University. Ponds are randomly assigned to receive one of two treatments: fertilizer added or control (no fertilizer). Once a month during the period April through October, 10 5-liter water samples are take from random locations in each pond. Algae biomass is measured on three one-liter random subsamples taken from each water sample.

   Variable of Interest: _____

   Population of Interest: _____

   Replicate: _____

   Pseudo-replication? Yes / No   Lack of Independence?   Yes / No

   Explanation: _____

   _____

   _____

c. The experiment described in part (b) was done to address the question about the effect of fertilizer runoff on lakes, but it was actually implemented using small artificial ponds. Describe the advantages and disadvantages of doing the study with artificial ponds rather than using real lakes.

21. a. *Study Question:* Do ambient levels of ozone reduce lung capacity in humans? Ozone is an air pollutant that may cause irritation to the lining of the lungs, resulting in constriction of bronchial passages and fluid deposition in the lungs. *Study Design:* The investigator obtains 50 volunteers, who are randomly assigned to a control or a treatment group. The groups meet at noon on a summer day in an open-air park in downtown Los Angeles. The treatment group wears breathing masks that provide them with air that has been charcoal-filtered to remove ozone normally present in the Los Angeles air. The control group wears similar masks, but receives air that has not been filtered. After five hours, the lung capacity of each individual is remeasured. This procedure is repeated with the same people on five successive days.

Variable of Interest: _____

Population of Interest: _____

Replicate: _____

Pseudo-replication? Yes / No   Lack of Independence?   Yes / No

Explanation: _____

_____

_____

b. *Study Question:* What is the prevalence of lead poisoning among children in Baltimore? Lead poisoning is a common problem in older eastern cities where many homes were built before a ban on lead-based interior house paint in 1970. This old paint degrades, resulting in paint dust and flakes falling off the walls and onto home surfaces, where children may inhale or swallow the paint. Newer homes do not contain lead-based paint. *Study Design:* The investigators go to *every* elementary school in the Baltimore public school system, randomly select 10% of the children at each school, and obtain permission to take a blood sample to determine lead concentration.

Variable of Interest: _____

Population of Interest: _____

Replicate: _____

Pseudo-replication? Yes / No   Lack of Independence?   Yes / No

Explanation: _____

_____

_____

c. *Study Question:* What is the prevalence of obesity among the people of Carmel, Indiana? *Study Design:* This town is comprised almost entirely of single family houses. Investigators randomly select 100 street addresses and obtain height and weight data (used to identify obesity) for every person living at each address.

Variable of Interest: _____

Population of Interest: _____

Replicate: _____

Pseudo-replication? Yes / No   Lack of Independence?   Yes / No

Explanation: _____

_____

_____

22. An investigator is conducting an experiment to determine the effect of fertilization on corn yield. However, the field that is available for this study is known to have two different areas that differ in soil type, such that water rapidly drains from one of the areas but tends to be retained in the other. Diagram an appropriate study design for dealing with this variation in soil within the experiment to determine the effects of fertilization. Use the study design diagrams in the text as examples, but replace the general terms with terms specific to this question. *Note:* In agricultural yield studies, a replicate is usually a plot of ground for which total crop yield is determined.

23. An investigator wants to determine the if taking calcium supplements can increase bone density in post-menopausal women. There is wide variation in bone density among women due to ethnic/racial differences, health/exercise history, and other factors. The investigator expects that the effect of the calcium supplements will likely be small. What type of study design would be most likely to detect any effect of the calcium supplements, given the wide range of variation among the study subjects? Explain.

24. An investigator believes that the effect of atmospheric ozone on plant growth may depend on water availability. Adverse effects of ozone occur when this gas is taken up by the leaf. Under conditions of low water availability, the openings in the leaf where gas exchange occurs close, preventing water loss from the leaf and uptake of gases from the air. Diagram an experiment to test for this interactive effect of ozone and water availability on plant growth using two levels of ozone (high, low) and two levels of water (wet, dry). Use the study design diagrams in the text as examples, but replace the general terms with terms specific to this question. *Note:* In this type of physiological ecology study, the replicate is usually an individual plant.

25. Researchers want to determine the effect of atmospheric ozone on the growth of loblolly pine (one of the most important timber producing tree species in the United States). Two types of studies were conducted.

Experimental Study: Five 3-yr-old pine saplings are randomly assigned to each of 30 open-top chambers. (The top is open to receive ambient rainfall and solar radiation but closed on the sides and bottom to allow for control of atmospheric gases in the chamber.) The saplings are spaced sufficiently far apart that they do not interact (e.g., compete for light, water, or nutrients). Fifteen chambers are randomly assigned to each of two treatment groups. Group (1) receives ambient air that contains ozone, and group (2) receives charcoal-filtered air (ozone removed). All other environmental conditions are kept constant among all chambers. Height of all pine saplings is measured at the beginning of the experiment and on Sept. 1 in each of the next three years. At the end of the experiment, the saplings are harvested, total above ground biomass (kg) is measured, and average height (for each of 3 years) and final biomass of trees in the two treatment groups are compared.

Natural Experiment: Atmospheric ozone concentrations downwind of Atlanta are much higher than at locations upwind of Atlanta. Researchers randomly select ten 30-yr-old loblolly pine plantations located upwind of Atlanta and five same-aged loblolly pine plantations downwind of Atlanta. Ten trees randomly selected in each plantation are harvested, and their total above ground biomass (kg) is determined. The average biomass of trees located up- vs. downwind of Atlanta is compared.

   a. What are the advantages of this natural experiment approach over the controlled experiment using saplings grown in open-topped chambers?

b. Describe potential problems for using the natural experiment approach to infer the effect of ozone of loblolly pine growth. Give examples when appropriate.

c. Do either one of these study designs suffer from pseudo-replication or any other problem that would violate the assumption of independent observations? Explain.

26. For each of the following situations (a) through (h), select the most appropriate sampling or experimental study design from the list below. Explain your choices.

Completely random sampling	Completely randomized experiment
Systematic sampling	Matched-pairs experiment
Stratified sampling	Block design experiment
Natural experiment	Factorial design experiment

a. A wildlife biologist wants to obtain an estimate of the abundance of northern spotted owl within a national forest. Based on prior research and experience, the biologist knows that the owls are more abundant in mature forests (> 200 yrs old) than in younger forests (between 20 and 200 yrs old) or areas recently logged (< 20 yrs). Based on an analysis of maps, 20% of the national forest is in mature forest, 50% in young forest, and 30% has been recently logged.

b. A student in a wildlife biology class wants to determine the abundance of a rodent species in a local nature preserve. She is concerned that she only has a modest number of traps and time to do the project, but she wants to make certain that she obtains data from all parts of the preserve.

c. Another student in the same wildlife biology class also wants to determine the abundance of a rodent species in the same nature preserve. Based on information from the literature, he knows that the spatial distribution of this species is very regular, due to the fact that individuals establish and defend territories. The student is concerned that the sample estimates be unbiased.

d. A toxicologist wants to determine if a combination of two estrogen-like pesticides have a greater effect on reproductive development in lab rats than either one separately.

e. A toxicologist wants to determine if a new estrogen-like pesticide affects reproductive system development in mammals. He plans to expose laboratory rats to the pesticide but is concerned that the pesticide may have differential effects on male rats than on female rats.

f. Some toxicologists criticize experiments on the effects of estrogen-like pesticides that use rats because rats have a different estrogen physiology than that of humans. These toxicologists want to obtain data on the effects on humans of exposure to estrogen-like pesticides, singly and in combination. How could these data be obtained without a serious breach of ethics?

g. A health and nutrition researcher wants to determine if people who are on a calorie-restriction diet and do an hour of aerobic exercise three days per week lose more weight than people who just restrict caloric intake, people who just exercise, and people who do neither.

h. A public health researcher wants to determine the average blood lead concentration for children in elementary school (grades K through 5), but has reason to believe the concentration may differ among neighborhoods where predominantly lower-, middle-, and upper-socioeconomic class populations live.

27. *Identifying Pseudo-Replication:* For each of the following study descriptions, identify the "replicate" unit appropriate for the study question. If you identify pseudo-replication, describe how the multiple measurements on a replicate should be analyzed. If there is a lack of independence, describe how the measurements are related.

a. *Study Question:* Do neurons in mouse brains develop more synapses when the mice live in complex environments than when the mice live in simple environments? *Study Design:* A female mouse and her litter of eight newborn pups are placed in a cage with many colored hiding boxes, a running wheel, and various "toys." A second female mouse and her litter of eight pups are placed in a bare cage. When the pups are 6 months old, they are sacrificed and 10 thin sections are obtained from each mouse brain. On each section, the number of synapses on a random sample of 15 neurons are counted. *Note:* The real question of interest might be how environmental stimulation affects brain development in children. Since doing this experiment with children is unethical, an experiment with mice, rats, or monkeys is used to study how mammalian brain development responds to environmental stimulation.

Replicate: _____

Pseudo-replication? Yes / No   Lack of Independence? Yes / No

Explanation: _____

_____

_____

_____

b. *Study Question:* Can corn yield per acre be increased by increasing the planting density (i.e., planting seeds at 4-inch intervals rather than 6-inch intervals)? *Study Design:* A field at an agricultural research station is divided into two halves. In one half, 20 separate 100 × 100 ft plots are planted at 6-inch intervals. In the other half, 20 separate plots of similar size are planted at 4-inch intervals. The yield of corn from each plot (bushels) is measured.

Replicate: _____

Pseudo-replication? Yes / No   Lack of Independence? Yes / No

Explanation: _____

_____

_____

_____

28. Research has shown that moderate wine consumption is associated with lower incidence of coronary heart disease, and that this is due to nonalcohol constituents in wine (especially polyphenols). Studies have shown that polyphenols reduced production of Endothelin-1 (ET-1), a chemical produced by cells in the walls of blood vessels that causes them to constrict, contributing to heart disease. Now researchers want to determine if different kinds of wine differ in their effectiveness for suppressing ET-1 production. The "study subjects" were tissue cultures of cow blood vessels. The investigators will apply alcohol-free extracts from each of three types of wine (Cabernet, Rose, and Chardonnay) to 30 tissue cultures (total of 90 cultures), and measure ET-1 concentration after one hour. (Based on a study by R. Corder et al., in the December 2001 issue of *Nature*.)

  a. What kind of experiment was this (completely randomized, block, or factorial)?

  b. Diagram this experiment, using the same format as shown in the text.

29. In the study described in Question 28 above, Carbernet was shown to reduce ET-1 production, reducing constriction of arteries and so the risk of heart disease. However, caffeine in coffee and soft drinks can cause constriction of blood vessels, and many people who consume wine also consume caffeinated drinks. Suppose the investigators in Question 28 wanted to study the interactive effects of caffeine and Cabernet extract on ET-1 production in cow blood vessel tissue cultures.

  a. What kind of experimental design would be most appropriate?

  b. Diagram this experiment using the same sample size per group as in Question 28, with two treatment levels for each of caffeine and Cabernet treatments.

Study Problems for Chapter 1 (with Answer Key)

1. Use the random numbers table in your textbook to randomly assign the nine people in the list below to three treatment groups (1, 2, 3) of three people each. Use sequential four-digit random numbers starting on line 101 column B, and reading down the column.

   Alex   Bob   Carol   David   Eddy   Fred   Gus   Harry   Ivan

2. A public health official in a large urban area wants to determine the prevalence of HIV-positive people within her jurisdiction. She wants her estimate of the percentage of the population that is infected to be unbiased and precise.

   a. Identify potential *sources of bias* in this context and describe how a study design could minimize bias in the estimated prevalence of HIV-positive individuals.

   b. Describe the aspect of sampling study design that would be most important for *minimizing random sampling variation* in her estimate of the prevalence of HIV infection.

3. A physical fitness researcher performs an experiment to determine if exercising several times for short periods each day gives the same fitness benefits as a single long bout of exercise. His study subjects are overweight female college students. The subjects are randomly assigned to different treatment groups. At the beginning of the experiment, the average weight of subjects for each group is approximately the same. After 12 weeks on the exercise regime, the amount of weight lost by study subjects is measured.

a. Why did the investigator assign subjects to the different treatment groups in a manner that ensured the average weight for each group was approximately equal at the beginning of the study?

b. What aspects of experimental design were most important for attaining treatment groups that had similar average weight? Explain.

4. A wildlife biologist wants to locate sampling points in a forest where she will determine the abundance of an endangered song bird species by observation and listening for its song.

a. Describe the justification for locating these observation points *systematically* and describe under what conditions this would be advisable.

b. Describe the justification for locating the observation points *randomly* and describe under what conditions this would be advisable.

5. People who are infected with HIV are able to minimize the amount of active virus in their bloodstream by taking a "cocktail" of various drugs. Many of these people also take various vitamins and herbal supplements that are reputed to enhance health, vigor, and disease resistance. One supplement that is widely used is extract of garlic, usually in pill form. An AIDS researcher has noticed that some of his patients maintain fairly high viral loads (counts of the HIV in the blood) in spite of the antiviral drug

cocktail. He wonders if one or more of the nutritional supplements they take might be reducing the effectiveness of the drug cocktail. One of his experiments focuses on garlic supplements. He obtains 40 volunteers who are HIV positive. All take the same antiviral drug cocktail, and all currently take garlic supplements. Diagram a *completely randomized experiment* that uses these volunteers to determine if taking the garlic supplements results in higher counts of active HIV in the bloodstream.

6. Describe a *natural experiment* that could be used to determine whether or not children who live in a home with one or more adult smokers are more likely to develop asthma than children who live in a home with no smokers.

7. Identify whether or not the following study design would produce data that violates the independent-observations assumption. If so, explain the nature of the "lack of independence" and what might be done to improve the study design. *Study Question:* Does watching violent cartoons increase antisocial behavior in children? *Study Design:* One class of children at a daycare center watches an hour of action/adventure cartoons that contain segments of violence. Another class (the control group) watches an hour of *Sesame Street*. During the hour immediately after watching the programs, the two classes of children are observed and the total number of incidences of antisocial behavior is counted for each group. This procedure is repeated with these same classes for five days.

Replicate: _____

Pseudo-replication? Yes / No   Lack of Independence? Yes / No

Explanation: _____

_____

_____

_____

8. Refer to the experiment described in Study Problem 5. HIV patients who have been on the antiviral drug cocktail for several years often experience increased active viral counts in their blood as the virus develops resistance to the drugs. The investigator who is assessing whether or not garlic supplements reduce the effectiveness of the antiviral drugs wants to account for the different viral loads of patients who have been on the drug cocktail for more or less time. Among his 40 volunteers, 20 have been taking the drug cocktail for less than 2 years and 20 have been taking the drug cocktail for more than 2 years. Diagram a block design experiment to assess whether or not garlic supplements reduce the effectiveness of the antiviral drug cocktail for HIV patients in these two "time-on-drugs" categories.

Answer Key for Chapter 1 Study Problems

1.

Name	RN	Ranked by RN		Treatment Group
Alex	3969	Ivan	0545	1
Bob	3703	David	0850	1
Carol	4010	Fred	1119	1
David	0850	Eddy	2555	2
Eddy	2555	Gus	2708	2
Fred	1119	Harry	3007	2
Gus	2708	Bob	3703	3
Harry	3007	Alex	3969	3
Ivan	0545	Carol	4010	3

RN = Random number.

2. a. *Sources of bias* include nonrandom sampling such that individuals are not representative of the entire population and consistent measurement error. In this context, bias could occur when randomly chosen people refuse to participate. Elderly people confident in their monogamous relationships may refuse to allow an intrusive or inconvenient blood test. People who know they are HIV positive may refuse because they are fearful of discrimination. If the sampling is done simply as a phone survey (*Question:* Are you HIV positive?), bias could occur due to ignorance or lying about HIV status, resulting in invalid measurement of prevalence. A sampling study that uses the most accurate blood test and maximizes participation by addressing concerns about convenience and discrimination would minimize bias (e.g., completely confidential, in-home blood tests).

b. *Minimizing random sampling variation* is accomplished by obtaining data for a large number of individuals.

3. a. Any difference in weight loss between the treatment groups *after* the treatment will be attributed to the effect of the treatment. This interpretation is valid only if the groups were similar *before* the treatment was imposed.

b. Equivalent treatment groups are attained by randomly assigning an adequate number of study subjects to each group. Even though individuals may differ, the averages for the treatment groups are more likely to be similar before treatment if randomization and replication are used to form the groups.

4. a. A systematic sampling design would be advisable if the investigator had limited resources and could only establish a small number of observation points. Systematic sampling would ensure even coverage of the study area. Systematic sampling would also be more efficient (most data for available time and resources) if travel between observation points was difficult.

b. Random sampling would be advisable if the investigator had reason to believe that the birds would be evenly distributed in the study area (for example, when breeding pairs establish territories). When the population of interest is distributed in a systematic manner, systematic sampling can result in biased estimates.

5.

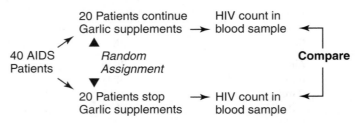

Completely Randomized Experiment

6. There are many possible approaches for doing this *natural experiment*. One approach is to do a phone survey of randomly selected households. Ask the respondents: (1) Does anyone in the household smoke? (2) Do any children in the household have asthma? Compute the proportion of Smoker households that have an asthmatic child and compare this to the proportion of Nonsmoker households that have an asthmatic child.

7. *Replicate:* The class of students (a group of children). Observations of individual children suffer from *lack of independence*. To observe a child exhibiting antisocial behavior, the child must be in a social group. The antisocial behavior of a single child in one of these classes would not be independent of the behavior of other children in the class. When one child exhibits antisocial behavior, the behavior of other children will often change. Hence, only the combined observation of all antisocial behavior in a class represents a single, independent observation.

Pseudo-Replication: Yes. Multiple observations of the same two classes of children over five days constitute repeated measurements on the same replicates. The investigator should total or average data for multiple days to obtain a single number for each class, and ideally, obtain observations for many more classes of students.

8.

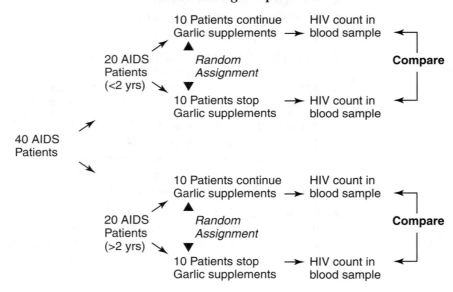

Block Design Experiment

Exercise A: Field Sampling Design

Objective Given a description and map of an area from which a random sample is to be taken, you will be able to implement a sampling design such that sample points are randomly located within the area.

Introduction There are a number of reasons for wanting to collect information on the characteristics of vegetation in a specified area (such as economic questions, plant ecology, and animal habitat analysis). It is usually the case that it is impossible to study the vegetation across the entire area of interest. Hence, it is necessary to obtain data from a representative sample of the area of interest and to use this sample data to describe the entire area. To implement a truly random sampling design, one must have a map of the area of interest with reference points that can be located without error. Random sample points are first located on the map using a random numbers table and either a grid or scaled measurements from mapped reference points. Based on these mapped sample points, compass bearings and distances from reference points are determined and these provide directions for locating random sampling points in the field.

Directions In this exercise you will randomly locate five sampling points with a map of Christy Woods (see Figure 1.1). By the end of the exercise, you should have the five points marked on the map and explicit directions for actually locating these plots in Christy Woods by measuring distances from a fixed geographical reference point. These sample points will represent the centers for five circular sample plots of radius 15 meters.

1. The first step is to establish a frame of reference that will allow you to actually locate randomly chosen points in the sampling area. This requires a geographic reference point that is both marked on a map *and* easily identified and located in the field. Road intersections, fence corner posts, USGS survey markers, and so on could all serve this purpose.

2. Given the location of this reference point, the next step is to specify rules regarding which range of values for random numbers may be used. For example, suppose you wanted to use an *x-y* grid to randomly locate points on a rectangular map area with length 150 mm and width 90 mm, and you planned to use a fence post that defines one corner of the rectangle as the reference point. For each random point on the map, you would first select a three-digit number between 0 and 150 for the *x*-value and then a two-digit number between 0 and 90 for the *y*-value. Each pair of random *x*- and *y*-values would describe the position of a randomly located point on the map. After you have located sampling locations on the map, you will need to use the map scale to convert the map distances (mm) to actual distances in the field (meters).

3. Take boundaries into account in locating the sample plots. Plots in an area with distinct boundaries (like a boundary between a cornfield and a woods) may be too close to the edge and may not be entirely within the area of interest. For example, if the plot radius is 15 meters, and the plot center falls 5 m from an edge of the study area, the plot will include area outside the woods. To prevent this, "buffer zone" rules are established so that the plot area will always fall within the area of interest. *Note:* The Christy Woods map has a 15-m buffer zone indicated by cross-hatching. If your random procedure for locating plots places a plot *center* within the buffer zone, that plot center should be rejected and a new set of random numbers used to relocate the plot.

4. Have spaced-out sample plots. When the number of plots is small, the plots should *not* be clumped together in a small part of the study area, leaving other parts of the area totally unsampled. One approach for avoiding this is to stipulate *buffer zones around plots*, such that one sample point may not be within a certain radius of another sample point. For this exercise, we will stipulate that sample points for plots with a 15-m radius should not be closer than 40 m to any other sample point. If two points are too close, drop the one last chosen and select a new random location.

5. Make sure all parts of the study area are sampled. With oddly shaped areas, it may be necessary to establish rules for the range of acceptable random numbers that might allow some points to fall outside the study area to ensure that all parts within the area are to be eligible for sampling. If a randomly located point falls outside the study area or within a buffer zone, simply delete that sample point and choose another pair of random numbers.

6. Once you have established ranges of acceptable random numbers for *x-y* map coordinates, use the random numbers table to obtain pairs of random numbers that represent distances from an easily located permanent geographic reference point, such as a fence corner post.

    a. Starting at a randomly chosen point in the table, select sequential random numbers that are within the range of acceptable numbers for the *x*-coordinate. If you are looking for three-digit numbers, you should always retain leading 0s in the three-digit numbers (e.g., 003), as this is the only way that small random number values can be obtained. Random numbers tables have clusters of four or five random numbers. If you only want three-digit numbers, you should read only the first three digits of each cluster of random numbers. When you get to the bottom of the first column, go back to the top of the next column. Continue on to sequential columns and repeat this process as necessary. Mark the position where you end this search after finding five appropriate *x*-coordinate numbers, as this will be the beginning point for the search for random numbers to position the sample points on the *y*-axis.

    b. For each of the *x*-axis numbers, continue searching in the random numbers table to select numbers that fall within the range of acceptable values for the *y*-coordinate. Write the *x*, *y* pairs of random numbers on the accompanying Worksheet for Exercise A.

Name _____ Date _____

7. Using the fence corner reference point labeled on the map as the origin of the *x-y* grid, locate the five sample points on the map using the random numbers as distances in millimeters. Mark each location with a dot and the plot identification number that corresponds to the number (1 to 5) where you recorded your random numbers on the worksheet.

8. Using the map scale, convert the map distances in millimeters to actual field distances in meters North and West from the fence corner geographic reference point. Write explicit field directions to locate the plot centers, including distances and compass directions.

Worksheet for Exercise A

1. Enter the Minimum and Maximum random number values that you will use to obtain random numbers from the table on the back of this page

Compass Direction	Axis	Range of Acceptable Values	
		Minimum	Maximum
E–W	X	_____	_____
S–N	Y	_____	_____

2. Implement the procedure using the following steps and the accompanying random numbers table on the back of this sheet.

   a. Randomly chose a starting point in the table and *circle that number*.

   b. Read down the column from your starting point and *underline those numbers that you actually use to randomly locate points in Christy Woods*. If you only need a two-digit number, underline only the first two digits of the four-digit cluster. You should always retain leading 0s in the multidigit numbers (e.g., 003), as this is the only way that small random number values can be obtained. *You must use numbers that fall within the range(s) specified above as you encounter them in the table.* When you get to the bottom of a column, go back to the top of the next column. Continue on to sequential columns and repeat this process as necessary.

   c. Mark the position where you end any part of your search, as this will be the beginning point for the search for random numbers for the next step in your procedure, if there are multiple steps.

   d. Enter the selected random numbers in the spaces provided in the table below. These designate the numbers you will use for each plot's center point.

3. Write directions for actually locating sample plots in Christy Woods. These should be compass directions (N, S, E, W) and distances, in meters, measured from the fence corner reference point marked on the map. Distances are computed from the random number *x*, *y* map coordinates using the map scale 1 mm = 1.7 m.

Plot Center	Random Number X	Random Number Y	Directions
1	_____	_____	_____
2	_____	_____	_____
3	_____	_____	_____
4	_____	_____	_____
5	_____	_____	_____
6	_____	_____	_____
7	_____	_____	_____
8	_____	_____	_____
9	_____	_____	_____
10	_____	_____	_____
11	_____	_____	_____

Random Numbers Table for Exercise A

Line	A	B	C	D	E	F	G	H	I	J	K	L
101	4947	3969	8489	3037	9647	7848	4673	4651	9815	9696	1325	9234
102	2910	3703	4783	4056	2928	0307	2597	3706	5040	5504	1506	6951
103	8929	4010	1386	6838	3648	2597	1965	7534	2586	7902	1762	9949
104	6064	0850	0782	6017	4211	0520	7601	5609	2202	2534	9263	2545
105	8719	2555	2358	5158	5489	5379	4362	9468	5227	3673	4335	4245
106	4551	1119	3068	6804	6233	2762	5297	0244	1509	0499	6120	4778
107	7110	2708	9821	6728	9212	0552	1532	0432	3032	2615	9962	3354
108	6806	3007	9199	8038	9679	1779	9495	5482	9896	0395	2134	7769
109	7363	0545	7662	8246	9328	5440	8612	8376	1763	3304	5030	6474
110	8354	8604	2828	9933	6325	5911	0096	7787	5390	6812	3458	4235
111	1960	3858	3988	3131	9570	9085	4957	7968	6398	6030	0145	9523
112	9569	7343	7451	5049	2794	0841	6848	5969	0863	1683	6088	9233
113	1349	0908	9871	0270	1635	0998	2891	9344	9889	1557	8880	3888
114	9413	9751	0157	2037	4299	3347	3089	4930	7778	8037	9615	3917
115	2699	9177	8106	2767	3942	9912	6458	6861	1903	9865	6528	5174
116	9197	1199	5380	9394	8399	5050	9503	9416	9944	3743	0971	9282
117	8646	4113	1953	6697	0377	2524	0168	7842	7577	8903	8852	1956
118	6322	5653	8647	9209	8746	6164	7868	7766	7264	5268	8183	9917
119	6046	7885	3893	9854	9046	2252	7753	1331	6265	3339	0832	4526
120	8034	9421	0330	0917	8837	9248	4805	5665	9484	1418	4124	7758
121	4722	8128	4377	0498	1023	7984	0912	6838	0293	3309	5417	3647
122	5174	4906	9394	5044	5762	2955	5529	9890	9944	7098	0258	5830
123	6218	6700	0570	6110	9102	2825	8063	3095	3704	3170	2274	9299
124	5924	4352	7146	6961	3954	5781	7598	2061	9327	7053	0393	2005
125	2546	8423	5111	2140	6543	2850	3092	6801	1008	0245	6541	9430
126	9563	3601	0990	0634	8940	9815	4664	2394	9411	9622	4432	2748
127	4936	9841	4854	0756	5827	5850	4105	1373	5937	3577	7574	8710
128	0211	5836	8171	2371	1092	4491	0224	3447	6817	0490	2121	6868
129	2587	7966	9624	9593	7404	0088	2176	5699	0190	2678	7696	7274
130	0583	5334	9529	4574	9374	0337	1368	3873	6759	6433	6800	9266
131	1417	9028	0964	0509	8441	3044	0098	2884	2184	6279	0950	9478
132	5737	1217	3248	9489	5095	2723	2528	2167	6797	9073	9152	4689
133	2814	7226	0886	8326	1576	0168	1133	9002	5807	1078	1848	2647
134	6827	1198	8606	2957	5658	5614	9233	0420	0582	8625	0954	1477
135	2381	3478	0620	9587	3663	3627	7761	1846	9163	0494	4121	5823
136	4848	5337	9722	8160	8308	5320	1854	8065	1611	7647	8829	5314
137	8694	4171	9364	1306	4527	1058	0431	9677	3328	3414	9467	8518
138	4390	9786	9313	9377	3885	4242	9355	5630	6973	4791	7990	2976
139	2046	9910	3038	9711	5710	4317	6312	9977	1094	2943	6305	9462
140	7387	0542	2437	5728	0573	4626	3044	8485	1102	8554	5996	6085
141	3198	4270	3028	9003	9226	8228	8041	8195	5722	5154	8521	5356
142	6865	7193	0245	4928	7649	2312	0326	0595	9526	1026	6500	4855
143	2131	2544	4940	0099	6303	9334	2471	6498	9694	7842	4222	9616
144	0696	6674	3735	7036	9234	0485	8444	8269	2615	1640	2997	1846
145	6794	0493	7347	9386	6142	4646	9429	2523	3459	9267	4480	1129

4. Location of Plot Centers on Map

    a. Put a point on the map of Christy Woods at each of the randomly located x, y positions where a plot center is located. Label each plot center point with its plot number (1 to 5).

    b. Use the map scale to convert the rule about 40-meter minimum distance between plots to a rule expressed in millimeters. Determine if any plot-center points on the map violate this rule. If so, put an X through the plot center you dropped and randomly locate another plot center point using the same procedure described in part 2 above.

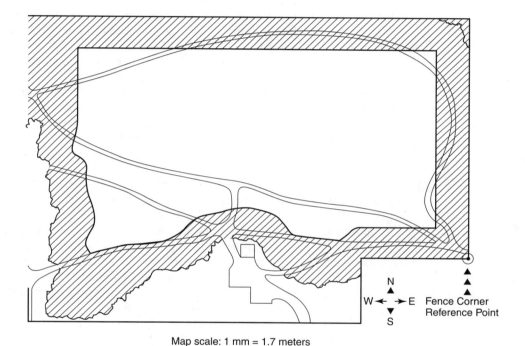

Map scale: 1 mm = 1.7 meters

FIGURE 1.1   Map of Christy Woods.   Cross-hatched area around the periphery of the woods is a buffer zone. No sample points should be located within or outside this part of the woods.

Exercise B: Random Sampling From Human Populations

Objective   Given a listing of a large human population (a University phone directory), you will be able to devise and implement a procedure for obtaining a random sample from this population.

Introduction   Researchers in many disciplines address questions that require making measurements or observations on samples of humans. Data are needed for public opinion polls, health surveys, and biomedical research. To obtain data for a "representative sample" of individuals, the researcher must use some form of randomized sampling to obtain a sample of people from the population of interest. There are a number of important ethical and statistical issues involved in any study of human subjects. These include full disclosure of possible adverse effects and the subject's voluntary participation. We will not go into these in detail here. You will devise a completely random procedure for selecting individuals from a University student body. In a real world study, these randomly chosen individuals would be asked to participate in a study. Those who agreed would comprise the sample. In this situation, "random sampling" implies that every individual in the population of interest is equally likely to be asked to participate in the study.

Directions   Carefully scan the University phone directory, taking note of how it is organized and any aspect of the directory structure that could be represented by numbers. Based on the organization and structure of the directory, write a brief description for how you could use a random numbers table to randomly locate an individual's name in the directory. Remember, you will have to actually implement this procedure, so make sure what you describe is doable with a reasonable amount of effort. *Hint:* You may specify a procedure that has more than one step, with a different random number for each step. You must specify what each random number represents, and the range of possible random numbers. For example, if I wanted to randomly select a number from 1 to 20, this would define the range of possible random numbers. You would ignore any numbers you encountered in the random numbers table that were not within this range.

Implement the procedure you described above to obtain 5 randomly chosen names from the University phone directory. For each name, write the random number(s) used to select that name and what the random numbers represent in the phone directory. The instructor should be able to read this description and locate that individual's name in the directory.

Worksheet for Exercise B

Description of Procedure. How will you use the random numbers table to obtain a random selection of names from the phone directory? If your procedure has multiple steps, describe the range of appropriate random numbers for each step.

Selection of Random Numbers. Implement the procedure you described above using the following steps and the accompanying random numbers table on the back of this page.

(a) Randomly choose a start point in the table and *circle that number*. (b) Read down the column from your starting point and *underline those numbers that you actually use to randomly locate names in the directory*. You do *not* need to use all digits in each four-digit random number cluster. If you only need a two-digit number, underline only the first two digits of the four-digit cluster. You should always retain leading 0s in the multi-digit numbers (e.g., 003), as this is the only way that small random number values can be obtained. *You must use numbers that fall within the range(s) specified above as you encounter them in the table.* When you get to the bottom of a column, go back to the top of the next column. Continue on to sequential columns and repeat this process as necessary. Mark the position where you end any part of your search, as this will be the beginning point for the search for random numbers for the next step of your procedure, if there are multiple steps. Enter the selected random number(s) in the space after the randomly selected name from the directory. After each name and set of random numbers, describe how those numbers represent a random position (name) within the phone directory.

	Individual's Name	Random Number(s)	What Does the Number Represent?
1	_____	_____	_____
2	_____	_____	_____
3	_____	_____	_____
4	_____	_____	_____
5	_____	_____	_____

Random Numbers Table for Exercise B

Line	A	B	C	D	E	F	G	H	I	J	K	L
101	4947	3969	8489	3037	9647	7848	4673	4651	9815	9696	1325	9234
102	2910	3703	4783	4056	2928	0307	2597	3706	5040	5504	1506	6951
103	8929	4010	1386	6838	3648	2597	1965	7534	2586	7902	1762	9949
104	6064	0850	0782	6017	4211	0520	7601	5609	2202	2534	9263	2545
105	8719	2555	2358	5158	5489	5379	4362	9468	5227	3673	4335	4245
106	4551	1119	3068	6804	6233	2762	5297	0244	1509	0499	6120	4778
107	7110	2708	9821	6728	9212	0552	1532	0432	3032	2615	9962	3354
108	6806	3007	9199	8038	9679	1779	9495	5482	9896	0395	2134	7769
109	7363	0545	7662	8246	9328	5440	8612	8376	1763	3304	5030	6474
110	8354	8604	2828	9933	6325	5911	0096	7787	5390	6812	3458	4235
111	1960	3858	3988	3131	9570	9085	4957	7968	6398	6030	0145	9523
112	9569	7343	7451	5049	2794	0841	6848	5969	0863	1683	6088	9233
113	1349	0908	9871	0270	1635	0998	2891	9344	9889	1557	8880	3888
114	9413	9751	0157	2037	4299	3347	3089	4930	7778	8037	9615	3917
115	2699	9177	8106	2767	3942	9912	6458	6861	1903	9865	6528	5174
116	9197	1199	5380	9394	8399	5050	9503	9416	9944	3743	0971	9282
117	8646	4113	1953	6697	0377	2524	0168	7842	7577	8903	8852	1956
118	6322	5653	8647	9209	8746	6164	7868	7766	7264	5268	8183	9917
119	6046	7885	3893	9854	9046	2252	7753	1331	6265	3339	0832	4526
120	8034	9421	0330	0917	8837	9248	4805	5665	9484	1418	4124	7758
121	4722	8128	4377	0498	1023	7984	0912	6838	0293	3309	5417	3647
122	5174	4906	9394	5044	5762	2955	5529	9890	9944	7098	0258	5830
123	6218	6700	0570	6110	9102	2825	8063	3095	3704	3170	2274	9299
124	5924	4352	7146	6961	3954	5781	7598	2061	9327	7053	0393	2005
125	2546	8423	5111	2140	6543	2850	3092	6801	1008	0245	6541	9430
126	9563	3601	0990	0634	8940	9815	4664	2394	9411	9622	4432	2748
127	4936	9841	4854	0756	5827	5850	4105	1373	5937	3577	7574	8710
128	0211	5836	8171	2371	1092	4491	0224	3447	6817	0490	2121	6868
129	2587	7966	9624	9593	7404	0088	2176	5699	0190	2678	7696	7274
130	0583	5334	9529	4574	9374	0337	1368	3873	6759	6433	6800	9266
131	1417	9028	0964	0509	8441	3044	0098	2884	2184	6279	0950	9478
132	5737	1217	3248	9489	5095	2723	2528	2167	6797	9073	9152	4689
133	2814	7226	0886	8326	1576	0168	1133	9002	5807	1078	1848	2647
134	6827	1198	8606	2957	5658	5614	9233	0420	0582	8625	0954	1477
135	2381	3478	0620	9587	3663	3627	7761	1846	9163	0494	4121	5823
136	4848	5337	9722	8160	8308	5320	1854	8065	1611	7647	8829	5314
137	8694	4171	9364	1306	4527	1058	0431	9677	3328	3414	9467	8518
138	4390	9786	9313	9377	3885	4242	9355	5630	6973	4791	7990	2976
139	2046	9910	3038	9711	5710	4317	6312	9977	1094	2943	6305	9462
140	7387	0542	2437	5728	0573	4626	3044	8485	1102	8554	5996	6085
141	3198	4270	3028	9003	9226	8228	8041	8195	5722	5154	8521	5356
142	6865	7193	0245	4928	7649	2312	0326	0595	9526	1026	6500	4855
143	2131	2544	4940	0099	6303	9334	2471	6498	9694	7842	4222	9616
144	0696	6674	3735	7036	9234	0485	8444	8269	2615	1640	2997	1846
145	6794	0493	7347	9386	6142	4646	9429	2523	3459	9267	4480	1129

Exercise C: Experimental Design

Objective   Given a research question, you and your classmates will be able to create an experimental study design, and implement it.

Introduction   Suppose you work for a company that is developing a new drug to treat bradycardia, a debilitating condition that manifests as a low heart rate, resulting in chronic fatigue. There is already an existing drug on the market (produced by a competing company) that provides partial relief by slightly increasing patient heart rate. For FDA approval you must demonstrate that your company's drug is effective for increasing heart rate. To out-compete the other company, you must demonstrate that your company's drug is more effective than the existing drug. Patients with this rare heart condition are scarce, and the treatment is extremely expensive, so sample size for the experiment is limited.

The Simulation   The effect of the two drugs on heart rate will be simulated by having the study subjects perform a very light exercise, standing and sitting repeatedly, moving up and down about once every second. The existing, "partially effective" drug will be simulated by performing five stand-and-sit repetitions. The new drug will be simulated by performing 15 stand-and-sit repetitions. Heart rate, measured by pulse counts, will be used to assess the effects of the two drugs.

Create an Experimental Design   The class will be divided into two groups of approximately equal numbers of students. This is intended to simulate equal resources for performing the study. One group should develop a completely randomized experimental design to test if the new drug is more effective than the existing drug. The second group should develop a before-after or matched-pairs design to address the same question. Both groups should develop a before-after or detailed experimental design and protocol that will provide the most unambiguous data regarding which of the two drugs is the better treatment for bradycardia. Any differences in the response of patients after the treatment(s) should be attributable solely to differences in the treatments, without the influence of other factors. Answer the questions below to organize your discussion about study design.

    a. How will you measure the responses of the subjects to the treatments?

    b. What are the sources of random variation and measurement error that might affect your measurements of response to the treatments?

c. How will you minimize random variation and measurement error?

d. What are the possible sources of bias in your experiment?

e. How will you use randomization to reduce or eliminate bias?

f. What else will you do to minimize or eliminate bias?

g. Draw a diagram of your experimental design.

**Implement the
Study Design**

1. Perform your experimental design using the members of your group for the simulation.

2. Record your data on the accompanying Data Sheet so that it is clear which data values are associated with each treatment and with each subject (each of whom can be identified by name or number).

3. Give your data sheet to the instructor before you leave class.

Data Sheet

Type of Experimental Design: _____

Note: If you use a matched-pairs design, list the two names of paired subjects one after the other and connect with a bracket. If you use a before-after design, list the subject's name on the first line along with the pulse rate for the 5 stand-sit treatment. List the pulse rate for the 15 stand-sit treatment on the next line, with no entry under Subject Name.

Subject Name (#)	Treatment (5 or 15)	Pulse Rate (beats per minute)
_____	_____	_____
_____	_____	_____
_____	_____	_____
_____	_____	_____
_____	_____	_____
_____	_____	_____
_____	_____	_____
_____	_____	_____
_____	_____	_____
_____	_____	_____
_____	_____	_____
_____	_____	_____
_____	_____	_____
_____	_____	_____
_____	_____	_____
_____	_____	_____
_____	_____	_____
_____	_____	_____
_____	_____	_____
_____	_____	_____
_____	_____	_____

2 Exploratory Data Analysis: Using Graphs and Statistics to Understand Data

Homework Problems

1. Bioengineers have developed a substance that stimulates the natural production of growth hormone of livestock animals so that they grow faster and bigger. To test the efficacy of the new bioengineered substance, 40 one-day-old female chicks of the Rhode Island Red breed, obtained from a single supplier, were randomly assigned to two groups of 20 chicks each. The control group was injected with the test substance after it had been inactivated by heat, while the treatment group was injected with the active substance. After 30 days, the following chicken weights (in grams) were measured:

   Control

   390 391 386 376 313 369 412 402 376 420 389 409 370 394 406 302 365 395 380 418

   Treatment

   382 418 402 396 435 404 394 427 407 309 418 420 428 421 448 393 431 380 411 387

   a. Why do you think that the investigators used only female chicks of the same breed, obtained from the same supplier?

   b. Make a stem-leaf plot of the data for each of the two groups.

     c. In each group, what is the frequency of data values that are at least 400?

       Control: _____   Treatment: _____

     d. In each group, what is the relative frequency of data values that are at least 400?

       Control: _____   Treatment: _____

2. Log-on to the following web site, **http://www.stat.sc.edu/rsrch/gasp/**, and scroll down to Educational Procedures. Select **A Histogram Applet** from the items in this list. Read the description of the data represented in the histogram. Below the histogram is a scale with a black triangle. Modify the "bin width" of the histogram bars by dragging the triangle along the scale. Compare the distribution (center, shape, spread, gaps, outliers) when the bin width is narrow versus the distribution when the bin width is large. What do you conclude from your observations?

Name _____   Date _____

3. a. Construct a stem-leaf plot for the data values listed below. The leaves on each row should be sorted in numerical order.

38	32	39	45	32	42	50	48	36	67	52	44	35	23	63
55	46	30	20	13	12	23	34	43	50	65	72	80	91	77

b. With the help of this stem-leaf plot, determine the minimum, 25th percentile, median, 75th percentile, maximum, and interquartile range (IQR).

Maximum: _____   75th percentile: _____   Median: _____

25th percentile: _____   Minimum: _____   IQR: _____

c. Use the 1.5 × IQR criterion to determine if these data values include any outliers. Show the calculations you used to determine the upper and lower boundaries for identifying outliers.

d. Using a hand calculator, compute the mean of these data values and enter the value below.

Mean: _____

e. Both the mean and the median are summary statistics that describe the center of a distribution. Compare these two values and explain why they are different or the same, based on the nature of the distribution. Which of these two statistics best represents the center of the distribution? Explain.

    f. Sketch a modified box-and-whisker plot for these data. Your boxplot should include an appropriately scaled Y-axis (see examples in the text).

4. a. Data for the two groups of chicks described in Problem 1 are given below. Find the five-number summary of the data values (minimum, 25th percentile, median, 75th percentile, maximum) and the IQR. The easiest approach is to use a stem-leaf plot for each group, with the leaves in numerical order.

Control

390 391 386 376 313 369 412 402 376 420 389 409 370 394 406 302 365 395 380 418

Treatment

382 418 402 396 435 404 394 427 407 309 418 420 428 421 448 393 431 380 411 387

Name _____   Date _____

Control Stem-Leaf Plot   Treatment Stem-Leaf Plot

Control Group   **Treatment Group**

Maximum: _____   Maximum: _____

75th percentile: _____   75th percentile: _____

Median: _____   Median: _____

25th percentile: _____   25th percentile: _____

Minimum: _____   Minimum: _____

IQR: _____   IQR: _____

b. For each experimental group, determine if there are any outliers based on the $1.5 \times$ IQR criterion. Show the computations you used to identify outliers.

Name _____   Date _____

c. Sketch side-by-side modified boxplots for the data values in each of the two experimental groups, control and treatment. Plot both boxplots on a single *Y*-axis to facilitate comparisons between the groups (see examples in the text).

Modified Boxplot　　　　　　　**Modified Boxplot**
Control Group　　　　　　　　**Treatment Group**

d. Use a calculator to compute the mean for each of the two experimental groups.

　　　　　　Control Group　　　　　　**Treatment Group**

Mean　　_____　　_____

5. Enter the data values listed in Problem 1 into a spreadsheet or statistics program work-sheet. (See Tutorial 3 for a description of how to enter data.)

a. Produce a separate histogram of data values for each of the two groups of chicks. If possible, make the X-axis scales of these two histograms similar. (See Tutorial 2 for a description of how to perform graphical and statistical exploratory data analyses.) Copy/Paste both these histograms onto a single page in a word processor document (see Tutorial 1). This is intended to facilitate comparison of these histograms and help you identify any difference that might reflect an effect of the bioengineered substance on chicken growth. Try arranging the histograms side-by-side and over-under on the page to see which arrangement makes it easier for you to make comparisons.

b. Make a side-by-side modified boxplot for the control and treatment groups (see Tutorial 2). Copy/paste this graph to a second word processor page.

c. Produce a summary statistics print-out that includes Group Name, Count (sample size), Mean, Median, Standard Deviation, IQR, Minimum, and Maximum (see Tutorial 2). Copy/Paste the summary statistics print-out and boxplot to the same word processor page as the boxplots. Append these two pages to this workbook page.

d. Histograms and side-by-side boxplots both display the distribution of data values for the two groups. Which of these graphics displays the details of the distributions more clearly? In which graphic can you more easily see the difference between the two data distributions? Explain.

e. The mean and median are both measures of the distribution center. Compare these two summary statistics for each group. Do they give the same answer regarding where the center is located? If not, why?

f. What measures of center and spread would be most appropriate for summarizing these two distributions of chick weights? Explain your answer using the words "resistant" and "sensitive."

g. Based on these graphical and statistical analyses of the data from the experiment described in Problem 1, what do you conclude regarding the effectiveness of the new bioengineered substance for enhancing the growth of chickens? Refer to specific graphics and/or summary statistics to support your conclusion.

6. An important assumption in the popular conception of the "balance of nature" is that predators control the abundance of their prey. If predators do control prey populations, when the predator population decreases, the prey population should increase and vice versa. That is, *there should be a negative association between predator and prey population sizes*. A study in Sweden used a disastrous decline in fox populations caused by a canine distemper epidemic (1981 to 1990) to determine if any of the fox's prey populations increased.

Fox population size was measured based on the number of fox litters found. Foxes are very secretive, and it was easier to find their dens with kits than to count adult animals. Black grouse population size was estimated based on the number of males observed. Again, these animals are secretive, except during mating season when males put on very visible and vocal displays to attract females. The population size of voles (small mouse-like animals) was estimated based on the number of animals trapped with a fixed number of traps set out over a fixed number of days. None of these variables actually quantify population size. However, if these variables are measured in exactly the same way every year, variation over time in these variables is a valid measure of variation in population size over time. The resulting data are given below. The years in bold type indicate the period of the distemper epidemic.

Year	Fox Litters	Black Grouse	Voles
1973	9	no data	5.5
1974	9	60	2.1
1975	5	54	0.1
1976	8	52	1.2
1977	9	45	2.4
1978	4	75	1.1
1979	9	55	0.2
1980	7	68	1.8
1981	4	72	0
1982	3	65	0
1983	4	42	0.4
1984	5	101	2
1985	4	96	0
1986	1	88	0
1987	5	92	1
1988	5	81	3
1989	3	70	0.4
1990	8	73	0.6
1991	7	45	0.2
1992	8	26	0

Enter these data into a spreadsheet or statistics program worksheet and create two scatterplots: Fox (*X*) vs. Grouse (*Y*) and Fox (*X*) vs. Voles (*Y*). See Tutorial 3 for data entry and Tutorial 2 for the way to make scatterplots. Copy both scatterplots to a single word processor page. Type a title on this page: **Chapter 2 Problem 6**. Below your scatterplots, you should *type answers* to the following questions.

a. *Based on the scatterplots,* describe the associations between fox abundance and the abundance for each of the two prey species. (Is the association positive or negative, weak or strong, linear or nonlinear?) What do you conclude about whether or not foxes control these two prey populations?

b. *Based on the timeplots in Figure 2.1,* state your conclusion as to whether or not each prey population size is negatively associated with fox population size. Explain your answers based on what you see in the *timeplots*.

c. Was it easier for you to see and describe the associations between fox abundance and prey species abundance in the scatterplots or in the timeplots? Explain your answer.

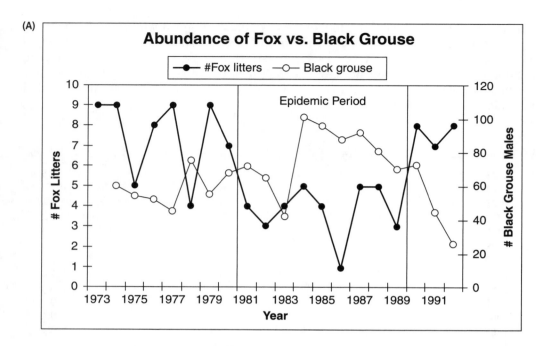

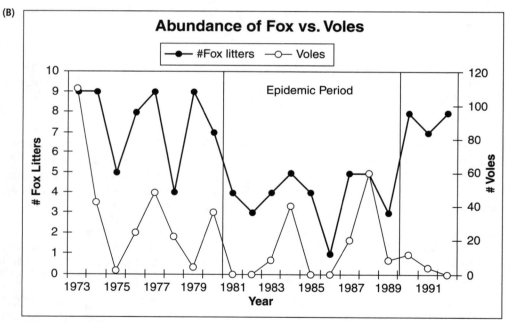

FIGURE 2.1 Timeplots.

7. For each of the following situations in parts (a)–(f), select the single best type of graph from the list below and briefly explain your answer.

Histogram	Side-by-side histogram	Scatterplot
Stem-leaf plot	Back-to-back stem-leaf plot	Timeplot
Boxplot	Side-by-side boxplot	

a. A public health official wants to determine if there is an association between the age of houses and the prevalence of lead poisoning in children. Lead was a common ingredient of interior house paint until it was banned around 1970. Paint flakes and dust from deteriorating paint in old houses are often consumed by small children. For each of 125 census tracts in her jurisdiction, the official determines the percentage of houses built before 1960 and the percentage of children who live in those tracts who were identified as having blood lead levels greater than 20 μg/dL during a school screening program for lead poisoning.

b. Suppose the public health official wanted to compare the distributions of blood lead concentration data among samples of children from four different city neighborhoods that differ in economic status. What graph type would be most appropriate if she wanted to compare the data distributions for the four neighborhoods with greatest clarity?

c. Suppose the public health researcher gets the data for the first 40 children in a sample from the lower-income neighborhood and she wants to do a quick summary of the results, but she does not have access to a computer. She also wants to compute a variety of summary statistics. What single graphic would facilitate this preliminary analysis?

d. Suppose the public health researcher *did* have access to a computer and wanted to create a graphic to use in a meeting of neighborhood residents to show the blood lead concentration results. She wants a graphic that the residents will be able to understand and that will look "professional."

e. Suppose the public health official has blood lead concentration data for 20 children who live in houses built before 1970 and for 20 children whose houses were built after 1970. She wants to compare the distributions of blood lead concentration between these two groups of children to assess if children in older houses have higher blood lead levels. What type of graph would be best if she had access to a computer and wanted a graphic that would be easily understood by the public and would have a "professional" appearance?

f. What type of graph would be best if the public health official in question e was doing a quick analysis to explore the data and did *not* have access to a computer?

Supplemental Problem

8. Although physical exercise has been recommended by some as a means for preventing or slowing loss of bone density in postmenopausal women, physical exercise may not be possible for some women. Some have proposed that much smaller strain on bones than is associated with physical exercise can have similar therapeutic effects on bone density. To test this hypothesis, investigators treated adult female sheep with a weak, high-frequency vibration to the hind limbs for periods of 20 min/day, 5 days/wk. Fourteen ewes were randomly assigned to this treatment group and 14 to a control group that did not receive this treatment. The two groups of sheep were kept together in a pasture except during the treatment periods. After one year of this treatment, density (gm cm^{-3}) of a bone in the left hind leg was measured. One ewe in the treatment group died due to causes other than the treatment. The data are presented below. (Based on a study by C. Rubin et al., published in *Nature* 412:603–604.)

Bone Density Data

Control	Treatment	Back-to-Back Stem-Leaf Plot
169	247	
216	230	
144	225	
204	213	
143	296	
184	280	
172	190	
160	253	
157	256	
173	227	
162	208	
154	221	
167	217	
133		

a. Construct a back-to-back stem-leaf plot of the data values for these two treatment groups in the space to the right of the data. The leaves on each row should be sorted in numerical order with the smaller leaves next to the column of stems.

b. Based on these two data distributions, determine the relative frequencies for each of the following ranges of data values for each group.

Bone Density	Control Group	Treatment Group
Values < 200	_____	_____
Values > 200	_____	_____

c. Using a hand calculator compute the mean for each group.

Control: Mean = _____

Treatment: Mean = _____

d. Using the stem-leaf plot, determine the five-number summary and interquartile range for each group. Then enter the values in the appropriate spaces below.

Control   **Treatment**

Maximum: _____   Maximum: _____

75th percentile: _____   75th percentile: _____

Median: _____   Median: _____

25th percentile: _____   25th percentile: _____

Minimum: _____   Minimum: _____

IQR: _____   IQR: _____

e. Perform the calculations to determine if any of the data values in either of the two groups are outliers, based on the $1.5 \times$ IQR criterion. List all outlier values for each group.

	Control Group	Treatment Group
Lower boundary value for outliers:	_____	_____
Upper boundary value for outliers:	_____	_____
Outliers:	_____	_____

f. Sketch side-by-side modified box-and-whisker plots for control and treatment group data that display any outlier values. Include a single, appropriately scaled *Y*-axis for these boxplots.

    g. Based on a comparison of the two data distributions, what do you conclude regarding the question of using low-strain–high-frequency vibration to increase bone density. Explain your answer based on the exploratory data analyses you performed.

Study Problems for Chapter 2 (with answer key)

1. Suppose you are given the following data values:

168	201	193	210	208	225	230	205	201	189	196
216	211	203	195	219	214	220	227	217	194	228

   Produce a stem-leaf plot for these data values. Leaves for each stem should be in order with the smallest value closest to the stem.

2. Use the stem-leaf plot to determine the five-number summary and the IQR.

   Maximum _____   75th percentile _____   Median _____

   25th percentile _____   Minimum _____   IQR _____

3. Identify any outliers (using the $1.5 \times$ IQR criterion). Show your calculations below.

4. Produce a modified boxplot with an appropriate Y-axis.

5. Use a hand calculator to compute the mean.

   Mean: _____

Answer Key for Chapter 2 Study Problems

1. Stem-leaf plot   **2. Five-number summary + IQR**

16 | 8	Minimum = **168**
17 |	
18 | 9	
19 | 345<u>6</u>	25th percentile = **196**
20 | 11358	Median = (208 + 210)/2 = **209**
21 | 01467<u>9</u>	75th percentile = **219**
22 | 0578	
23 | 0	Maximum = **230**

IQR = 219 − 196 = **23**

3. Outliers: **There are no outliers.**

Lower limit = 196 − 1.5(23) = 161.5

Upper limit = 219 + 1.5(23) = 253.5

4. Boxplot

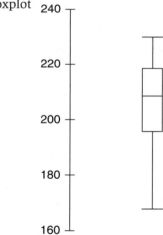

5. Mean = **207.73**

Exercise A: Exploratory Data Analysis

Objectives   You will be able to use a statistical analysis program to produce graphs and summary statistics that describe data distributions, and you will be able to use a word processor to create documents that include print-outs of results and text replies to questions. This exercise must be done in conjunction with Tutorials 1 and 2.

Introduction   The **Iris** data file contains data from a classic study of genetic variability in the dimensions of flower parts for three species of Iris. One objective of this study was to identify flower traits or combinations of traits that would allow for the correct assignment of individual plants to their respective species. For each species, individual plants exhibited greater or lesser amounts of variability in the length and width of their flower petals and sepals. For some traits, there was substantial overlap in the range of values among the three species. For other traits, there was much less overlap. You will apply procedures for displaying and describing data distributions to these variables (traits) to assess which traits are most useful for distinguishing different species.

Getting Started   Tutorial 1 describes the basics of operating your statistics computer program, including how to open the Iris data file, how to perform graphical and statistical analyses, and how to print the results. You should complete this tutorial before attempting to do this exercise.

Tutorial 2 contains explicit descriptions for performing specific graphical and statistical exploratory data analyses. This tutorial uses the Iris data file as a hands-on example for teaching you these procedures. Many of the graphs and print-outs you create for Tutorial 2 can be used for this exercise as well. Hence, you should complete Tutorial 2, then perform any additional analyses necessary to complete the assignment described below.

Graphical/ Statistical Analyses of the Iris Data   You will perform these analyses as you do Tutorial 2. You should Copy/Paste the analyses listed for each page onto a single word processor page. Each page should have the title **Chapter 2 Exercise A**.

Page 1:   Include four graphs of *side-by-side boxplots* (one for each flower-dimension variable). In each graph there should be three boxplots (one for each species). The complete species names should be listed along the *X*-axis.

Page 2:   *Summary statistics*: Include one table for each flower-dimension variable. There should be three lines of statistics, one for each species, with the species names listed in the first column of the print-out. Summary statistics should include: Group (species) Name, Count (n), Mean, Median, Standard Deviation, Interquartile Range, 25th percentile, 75th percentile, Minimum, and Maximum.

Page 3:   Include two *scatterplots*, one for petal length (*X*) vs. petal width (*Y*) and one for sepal length (*X*) vs. sepal width (*Y*). These scatterplots should contain the *combined data of all three species*.

Report   *Type answers* to the following questions. These answers should be on a separate page from the graphs and statistics pages described above. *Your answers must be in complete sentences and include an explanation.*

1. Which species had the largest sepals (the most showy flower)? Which species had the smallest sepals? Explain how you arrived at your answers, including which specific graphical and/or statistical data analyses were most useful to you in making your determinations.

2. **a.** Based on the side-by-side boxplots, which of the 12 (3 species × 4 flower dimensions) species and flower-dimension combinations (e.g., *I. setosa* sepal width) exhibited the greatest amount of variation among individual plants? Which exhibited the least amount of variation?

   b. Looking at the summary statistics print-outs, which one of the statistics listed best represents the amount of variation in a flower-dimension variable among individual plants of a single species? Using this summary statistic, which of the 12 species and flower-dimension combinations exhibited the most and the least amount of variation among individual plants?

   c. Which type of data summarization—graphical or statistical—made it easier for you to compare the amount of random variation in data values for species and flower-dimension combinations?

3. **a.** One of R. A. Fisher's objectives in the study that produced these data was to determine which variable or combination of variables was most useful in differentiating one Iris species from another. In this exercise, which of the four flower-dimension variables displays the greatest distinctions among the three species (least amount of overlap in data value)? That is, if you were going to use a single flower dimension to distinguish one species from the others, which one would you use?

   b. Explain how you arrived at your answer, including which graphical and/or statistical data analyses were most useful to you in making your determinations.

4. Based on the scatterplots that display the associations among pairs of flower dimensions, which of the two flower parts—petals or sepals—has the stronger association between their length and width. Explain what aspect(s) of the scatterplots lead you to this conclusion.

Exercise B: Data Analysis for Exercise C of Chapter 1

Objective You will be able to apply graphical and statistical analysis skills learned in Chapter 2 to the analysis of experimental data generated by before-after and completely randomized experimental designs implemented for Exercise C of Chapter 1.

Introduction In Exercise C of Chapter 1 you designed and implemented two different experimental designs to compare the effectiveness of a simulated new drug for bradycardia (a condition of abnormally low heart rate) versus a simulated existing drug for this condition. Assume you have collected the data from a before-after design and a completely randomized two-group design. The next step, as in a typical scientific research study, is to use graphical and statistical exploratory procedures to summarize the data. The purpose of this study was to compare heart rates of people given the new drug to heart rates for people taking the existing drug. To determine whether or not the new drug is better than the existing drug, we must summarize the numerous data values to obtain measures of center and spread. Graphics and statistics that display these characteristics allow us to better assess whether or not the new drug increases heart rate more than the existing drug.

Data Entry Before-after study design.

Data values should be entered in two columns (variables), titled **Tr5** and **Tr15** (indicating that the "existing" simulated treatment involved standing up and sitting down 5 times, and the "new" treatment involved standing up and sitting down 15 times). Data values for the Before and After measurements obtained from each individual must be entered on the same row in each variable. There must be the same number of Tr5 and Tr15 data values. See Tutorial 3 for instructions on data entry specific to the statistics software used in your course.

The data values that are most relevant for analyzing results from before-after study design are the difference values (computed as Tr15 − Tr5) for each individual in the sample. Statistics software has the ability to perform such calculations on the raw data values. You should consult Tutorial 2 for instructions on how to perform data transformations and calculations and compute a third column (variable) titled **Diff** that contains these difference values.

Completely randomized study design with independent treatment groups.

Data values should be entered in two columns (variables), titled **Tr5** and **Tr15**. There is no matching of data values on the same row, and data values for each variable can be entered in any order. It is possible that you will have a different number of data values in these two variables. See Tutorial 3 for instructions on data entry for the statistics software used in your course.

Some statistics software will require that the data be reformatted for some graphical or statistical analyses, such that all data values are in a single column (variable), and a second "Group" variable must be created to identify to which group each data value belongs. Most statistics software includes options for reformatting data back and forth between the formats described above. These options are usually found in a menu titled **Manip**. See Tutorial 2 for instructions on reformatting data.

Note: Some computer statistics programs will allow you to have the two different data sets (before-after and completely randomized) together on the same spreadsheet page.

Other programs may require that these two different data sets be on separate pages or even in separate data files.

Graphical Analyses

Create boxplots for assessing the difference in heart rate between the two treatments:

1. *Before-after design:* Produce a boxplot of the difference values (Diff = Tr15 − Tr5). If the two treatments are equally effective, this boxplot should be centered on a difference value of zero. If the new treatment Tr15 is more effective than the existing treatment Tr5, the center of the boxplot should be on a value greater than zero, and the range of the boxplot should indicate few values less than zero.

2. *Completely randomized design:* Produce side-by-side boxplots of Tr5 and Tr15 within the same graphic, on the same Y-axis. This will facilitate comparisons of the distributions of the two sets of data values.

   - Some statistics software will allow you to simply select these two variables and request a side-by-side boxplot in the Plot (Graph) menu.

   - Some statistics software will require that you "append" the heart rate data values for Tr5 and Tr15 into a single variable (named **Heart Rate**) and create a variable named **Group** that contains the categorical data values "Tr5" and "Tr15" to identify to which group each data value belongs. See Tutorial 2 for instructions on appending variables and creating Group variables. Once you have reformatted the data, specify **Group** as the group or *X*-variable and **Heart Rate** as the *Y*-variable.

Summary (Descriptive) Statistics

1. *Before-after design*: Select **Diff** and request the following Summary (or Descriptive) statistics: Count (sample size), Mean, Median, Standard Deviation, IQR, Minimum, Maximum. See Tutorial 2 for instructions on generating summary descriptive statistics specific to the statistics software used in your course.

2. *Completely randomized design*: Select **Tr5** and **Tr15** and request the same summary statistics listed above to obtain statistics for both groups.

Report

Copy/paste the graphics and summary statistics for data generated by each study design into separate pages in a word processor. Each page should have a boxplot and a listing of summary statistics. At the top of the page you should type a title in this format:

Data Analysis for < *enter Before-after or Completely Randomized* > **Study Design**

On each page, below the graphical and statistical summaries, type your assessment regarding whether or not the data indicate the new treatment Tr15 was more effective for increasing heart rate than the existing treatment Tr5. Explain your assessment, making explicit references to the graphical and/or statistical summaries. At this point in the course, your assessment will necessarily be somewhat subjective. The purpose of the exercise is for you to make a well-reasoned assessment based on the observed results, not to arrive at some "correct" answer.

In addition to your assessment of the difference between the two treatments, *type answers* to the following questions.

1. Which of the two study designs (before-after or completely randomized) provided clearer evidence of a difference between the two treatments? Explain why this design provided clearer evidence, based on topics discussed in Chapter 1.

2. Which form of data summarization—graphical or statistical—provided you with a better basis for making your assessment? Explain your answer.

3 Probability: How to Deal with Randomness

Homework Problems

1. Describe the sample space for the random variables described below. For discrete variables use the format {0, 1, 2, 3 ... < *maximum value* >}. For continuous variables use the format {0 to < *maximum value* >}.

   a. X = the amount of time per day that an organism is active (in hours).

   b. In the Physicians Health study, X = the number of physicians who suffered a heart attack out of 11,000 that received the aspirin treatment.

   c. X = the difference in the percentage of individuals who are clinically obese between a group that receives diet/exercise counseling and a control group that does not. (The difference is computed as treatment − control.)

   d. X = the proportion of students at a university who have blood type O−.

2. State whether or not the following probability statements or distributions are legitimate under the rules for probability. If not, explain why.

   a. When a coin is tossed, $P[\text{Heads}] = 0.55$ and $P[\text{Tails}] = 0.45$.

b. The probabilities that a person randomly selected from the U.S. population will fall into one of various racial classes are: $P[\text{Non-Hispanic Caucasian}] = 0.62$, $P[\text{Hispanic}] = 0.13$, $P[\text{African-American}] = 0.13$, $P[\text{American Indian}] = 0.009$, $P[\text{Asian}] = 0.04$, $P[\text{Other}] = 0.15$.

c. The probability of a beneficial mutation is very rare, approximately -0.9.

3. Two events are *disjoint* if it is not possible for a single observation (or individual) to fulfill the description of both events.

Two events are *independent* if the occurrence or nonoccurrence of one event is in no way related to or influenced by the occurrence of the other event.

For the following pairs of events, indicate if the events are disjoint and if the events are independent.

Event A	Event B	Disjoint? (Y/N)	Independent? (Y/N)
Subject is female.	Subject is a football player.	_____	_____
Subject is male.	Subject has blood type AB.	_____	_____
Subject is female.	Subject is obese.	_____	_____
Subject is male.	Subject is a rape victim.	_____	_____
Subject has blood type AB.	Subject has blood type O.	_____	_____
Subject is a binge drinker.	Subject does illegal drugs.	_____	_____
Organism is a reptile.	Organism is an amphibian.	_____	_____
Organism is a bird.	Organism breathes with gills.	_____	_____
Organism is warm-blooded.	Organism lays eggs.	_____	_____
Species is mammal.	Organism is listed as an endangered species.	_____	_____

4. Many games of chance are based on rolling two dice and determining X = the sum of the number of dots on the two faces on the upper surface. The matrix below displays all the possible outcomes in the sample space for tossing two standard dice.

Die #2

		1	2	3	4	5	6
	1	1,1	1,2	1,3	1,4	1,5	1,6
	2	2,1	2,2	2,3	2,4	2,5	2,6
Die #1	3	3,1	3,2	3,3	3,4	3,5	3,6
	4	4,1	4,2	4,3	4,4	4,5	4,6
	5	5,1	5,2	5,3	5,4	5,5	5,6
	6	6,1	6,2	6,3	6,4	6,5	6,6

Outcomes like 1,2 and 2,1 and all other similar pairs must be treated separately. There are 36 possible outcomes from tossing two six-sided dice.

a. Assuming fair dice, what is the probability associated with each of these 36 outcomes? Explain the logic behind your answer, based on the rules of probability.

b. In the game of Craps, the shooter wins by rolling a 7 or 11. What is the probability of getting one or the other of these outcomes on a single roll of two fair dice? Use the following steps to determine this probability.

(1) List all outcomes in the sample space of outcomes above that meet this description of 7 or 11.

(2) Based on your answer to part (a), what is $P[7 \text{ or } 11]$ on a single roll of two fair dice?

(3) Describe in words the logic of how you determined this probability.

c. In the game of Craps, the player can win progressively more money by shooting a 7 or 11 on successive rolls of the dice. Use the $P[7$ or $11]$ determined above to determine the probability that a shooter would roll a 7 or 11 on two successive rolls of a pair of fair dice. Show all work.

d. What is the probability that a shooter would roll a 7 or 11 on each of three successive rolls of a pair of fair dice? Show all work.

e. What assumption(s) are required for the probabilities you calculated in parts (c) and (d) to be valid?

5. Viral sexually transmitted diseases (STDs) are generally incurable and cause health problems that include mild discomfort, sterility, birth defects, and death. Viral STDs include HIV, herpes simplex-2, human papillomavirus (HPV), and hepatitis B. The following table provides estimates of the relative frequency of infected individuals in the U.S. population (obtained from the Centers for Disease Control and Prevention, year 2001 report). These relative frequency values can be used as estimates of probabilities that a randomly chosen sex partner would be infected with these diseases.

STD	Herpes	HPV	HIV	Hepatitis B
Prevalence	0.20	0.09	0.0016	0.0015

a. Use the simple addition rule to determine the probability that a randomly chosen sex partner would be infected with one or the other of these four STDs.

b. The simple addition rule requires that events be disjoint. Explain what "disjoint" means in the context of part (a) above. Assess whether or not the events (Has herpes), (HPV-infected), (HIV-infected), and (Has Hepatitis B) are disjoint.

c. Use the simple multiplication rule to determine the probability that a randomly chosen sex partner who has herpes would also be infected with the HIV virus.

d. The simple multiplication rule requires that events be independent. Explain what "independent" means in the context of part (b). Assess whether or not the events (Has herpes) and (HIV-infected) are independent.

6. The probability associated with a particular outcome for a random phenomenon is the *long-term* relative frequency of this outcome. Explain this relationship between probability and long-term relative frequency using the example of rolling a fair, six-sided die to determine the probability that the one-dot side will be face-up.

7. A survey of all nature preserves in the state of Indiana reports the relative frequencies of nature preserves in various size classes (acres) as listed below. These proportions were derived from a complete census of nature preserves, so the relative frequencies are also the probabilities that a randomly selected nature preserve will be of the indicated size.

Area (acres)	0–49	50–99	100–199	200–299	≥300
Probability	0.364	0.195	0.195	0.136	0.110

Let A be the event that a nature preserve is less than 100 ac, and let B be the event that the area is at least 300 ac.

a. Determine the probability for each of event A and event B. Show calculations.

$P[A]$ =

$P[B]$ =

b. Describe A^c in words and determine $P[A^c]$.

c. Describe in words the event (A or B). Determine the probability $P[A$ or $B]$.

d. Describe in words the event (A and B) and determine the probability $P[A$ and $B]$.

e. All these nature preserves were surveyed for the presence of an endangered bird species, and the proportion of preserves in which the bird was observed was 0.2. That is, the probability that this species would be present in an Indiana nature preserve is 0.2. Someone has proposed that this endangered species requires habitat areas of at least 100 ac, whereas others claim the presence or absence of this species is independent of area.

(1) If the presence of the bird is independent of habitat area, what proportion of the nature preserves would be both at least 100 ac *and* have this species of bird present?

(2) What might you conclude if the actual observed proportion of nature preserves in Indiana that meet these two criteria was 0.02?

8. A woman and a man (unrelated) each have two children. At least one of the woman's children is a boy. The man's oldest child is a boy. Given this information, is the probability that the woman has two sons equal to the probability that the man has two sons?

a. List all possible combinations of male and female children that fit the descriptions given for the children of the man and woman.

Possible combinations for woman's family:

Possible combinations for man's family:

b. You can assume gender of the children is determined at conception with an equal probability that a child will be male or female. Hence, all possible combinations that meet the descriptions above can be considered equally likely. On this basis, compute the probabilities for two events:

P[The man's children are both male] = _____

P[The woman's children are both male] = _____

Supplemental Problems

9. Suppose that the relative frequencies of the four ABO and Rh blood type groups in the U.S. population are as shown below. These relative frequencies correspond to probabilities that a randomly selected individual from this population would have one of these blood types.

	Blood Type				Rh Factor	
	A	**B**	**AB**	**O**	**Rh+**	**Rh−**
$P[\ ]$	0.4	0.1	0.05	0.45	0.85	0.15

Whether or not a person is Rh + or − is independent of that person's ABO blood type.

Each person has two genes, one on each of a pair of chromosomes, that code for that person's ABO blood type. The genotypes for individuals that have these blood types are listed below. (Each letter in the two-letter genotypes corresponds to one gene.)

Type A: AA or AO   **Type B**: BB or BO   **Type AB**: AB   **Type O**: OO

Likewise, each person has two genes on a pair of chromosomes that code for their Rh blood type. The genotypes that correspond to each Rh blood type are listed below.

Rh+: +/+ or +/−   **Rh−**: −/−

Each parent contributes one gene each for ABO and one gene for Rh to each of their children, and each of the two genes carried by a parent that code for ABO and Rh is equally likely to be passed on to each child.

a. What is the probability that a randomly chosen individual from the U.S. population would carry at least one gene for the B blood type? Show work.

b. What is the probability that a randomly chosen individual from the U.S. population would have blood type O− (universal donor)? Show work.

c. What is the probability that a randomly chosen individual from the U.S. population would have blood type AB + (universal receiver)? Show work.

d. Suppose both parents have blood type AB. What is the probability that a child produced by this couple would have:

Blood type A? _____

Blood type B? _____

Blood type AB? _____

e. Suppose a couple has blood types Rh+ and Rh−, but they don't know whether the person with Rh+ blood is +/+ or +/−. What is the probability that a child produced by this couple would have blood type Rh−?

10. A public health researcher wants to know if patients who come to the emergency room of a public hospital are more likely to be admitted to the hospital if they have health insurance (conversely, if patients who do not have health insurance are more likely to be treated in the emergency room and sent away, rather than admitted for more expensive treatment). She reviews the records for 1,000 emergency room patients and obtains the data presented below. Use these data to answer parts (a) through (e).

		Admitted to Hospital?	
		Yes	No
Insurance?	Yes	310	440
	No	65	185

a. Based on these data, compute the relative frequency of patients coming to this emergency room who are admitted to the hospital. This relative frequency is an empirical probability that a patient will be admitted. Show calculations.

b. Based on these data, compute the relative frequency of patients coming to this emergency room who do not have health insurance. Show calculations. This relative frequency is an empirical probability that a patient will not have health insurance.

c. Assuming that admission to the hospital is independent of whether or not a person has health insurance, use your results from parts (a) and (b) to compute the probability that a patient coming into the emergency room will not have insurance *and* will be admitted to the hospital anyway. Show calculations.

d. *Based on the data in the table above,* compute the relative frequency of patients coming to this emergency room who have no insurance but are admitted anyway. Compare this observed relative frequency to the probability computed in part (c). State your conclusion regarding whether or not admission to the hospital is independent of insurance status.

e. Using the general multiplication rule, compute the probability that a patient entering the emergency room will not have insurance and will be admitted.

11. The type of medical care a patient receives may be influenced by that patient's age. In a study of 2,000 women who had lumps in their breasts, the proportions of women who received a mammogram and lump biopsy were determined for women less than 65 years old and women 65 years old and older. The results in the table below are proportions of the total sample and number of women.

Age	Tests Done	Tests Not Done	Totals
< 65	0.321	0.124	890
	642	248	
≥ 65	0.365	0.190	1110
	730	380	
Totals	1372	628	2000

a. From the data presented in the table above, compute relative frequencies to estimate the probabilities for the following event descriptions. Show calculations to the side.

P[Patient age < 65] =

P[Patient received the test] =

P[Patient age ≥ 65] =

P[Patient did not get tested] =

b. Using these estimated probabilities, compute the probability of the event P[Patient ≥ 65 *and* Test Done]. Use the simple multiplication rule that assumes independent events. Show calculations.

c. Based on the data in the table above, what is the observed relative frequency of women who meet both criteria (Age ≥ 65) and (Test Done)? _____

d. Compare the probability computed using the simple multiplication rule to the observed relative frequency in the table. Under the multiplication rule $P[A$ and $B] = P[A] \times P[B]$ only if the two events are independent. What do you conclude? Explain.

e. Using the numbers of women in each class, compute the conditional probability that a woman would receive the test *if* she is 65 years old or older.

f. Use the general multiplication rule to compute the probability $P[A$ and $B]$ that a woman 65 years old or older will receive the test.

Study Problems for Chapter 3 (with answer key)

1. Describe the sample space for the following random phenomena.

  a. The number of students with brown eyes in a genetics class of 30 students.

  b. The temperature (°C) of distilled *liquid* water.

2. If a defective gene is located on the sex-determining chromosomes, one gender of child will be more likely than the other to manifest the associated genetic disorder. If the gene is not on the sex-determining chromosomes, then manifestation of the disease should be independent of gender. Suppose that the genetics of a couple is such that they have a probability $P = 0.25$ that a child born to them will have this disease. Assume the probability of a child being female is 0.5.

    a. What is the expected proportion of children that would be female and have the disorder if the genetic defect is not on the sex chromosomes? Show your work.

    b. Suppose that the defective gene is *not* on the sex chromosome. If this couple has two children, what is the probability that both will *not* have the disorder?

    c. If this couple has two children, what is the probability that *at least one* child will have this genetic disease? *Hint:* Use the complement rule.

3. Cystic fibrosis (CF) is a genetic disease that typically kills affected individuals by the time they are in their twenties. This disease is caused by a single defective gene. Humans have two copies of this gene. In people with CF, both copies of the gene are defective (genotype a/a). Some people carry one defective gene and one normal gene (genotype A/a); these people are "carriers," but do not develop CF symptoms. Each parent contributes one copy of this gene to each of their children, and each copy is equally likely to be passed on. Suppose a carrier man (genotype A/a) is married to a carrier woman (A/a).

    a. What is the probability that a child of this couple will have CF (a/a)?   _____

    b. What is the probability that a child will be a healthy noncarrier (A/A)?   _____

    c. What is the probability that a child will be a CF carrier (A/a)?   _____

Answer Key for Chapter 3 Study Problems

1. **a.** {1, 2, 3, 4, ... , 28, 29, 30}

   b. {0 to 100}

2. **a.** If the disease is not sex-linked (independent of gender), then

   P[female and disease] = P[female] × P[disease] = 0.5 × 0.25 = **0.125**

   b. P[first child healthy and second child healthy] = P[healthy] × P[healthy] = 0.75 × 0.75 = **0.5625**

   c. The event [At least one child with the disease] is the complement of the event [Neither child has the disease]. You computed the probability for the latter event above. Therefore:

   P[At least one child] = 1 − P[No child] = 1 − 0.5625 = **0.4375**

   Alternatively, this probability could be determined as follows:

   H = healthy   D = diseased

Possible outcomes for two children:	HH	HD	DH	DD
Probabilities:	0.5625	0.1875	0.1875	0.0625
	(0.75)(0.75)	(0.75)(0.25)	(0.25)(0.75)	(0.25)(0.25)

   P[At least one child with the disease] = P[HD or DH or DD] = 0.1875 + 0.1875 + 0.0625 = 0.4375

3. **a.** P[CF child (a/a)] = P[Father gives "a" gene *and* Mother gives "a" gene]

   = 0.5 × 0.5 = **0.25**

   b. P[Child noncarrier (A/A)] = P[Father gives "A" gene *and* Mother gives "A" gene]

   = 0.5 × 0.5 = **0.25**

   c. The sample space for the child includes three possible outcomes {Has CF, Carrier, Noncarrier}. Given the probabilities computed for parts (a) and (b) above, use the complement rule to determine the probability:

   P[Child is a carrier] = 1 − P[a/a] − P[Carrier A/A] = 1 − 0.25 − 0.25 = **0.5**

4 Developing Probability Distributions for Binomial Random Variables

Homework Problems

1. For each of the following situations, indicate whether a Binomial distribution is a reasonable probability distribution for the random variable X. If the distribution is not Binomial, *explain* which rule for Binomial variables was violated.

   a. You observe the sex of the next 50 children born at a local hospital; X = the number of girls among them.

   _____

   b. Ten ferrets are sampled from a population of 30 black-footed ferrets to determine the ratio of females in the population of this endangered species; X = the number of females in the sample.

   _____

   c. You want to determine the prevalence of lead poisoning in an inner city neighborhood. You randomly select 50 families who agree to participate, and you obtain blood samples from all individuals in these families. X = the number of individuals with blood lead levels that meet the standards for lead poisoning.

   _____

   d. To study the prevalence of lead poisoning among grade school children in a major urban area you randomly select 50 children from the combined rosters of all schools, totaling 18,000 students. X = the number of children out of 50 with blood lead levels that meet the standard for lead poisoning.

   _____

   e. You want to study the prevalence of a specific genetic marker among women whose mothers died of breast cancer (to determine if this could be a basis for assessing risk of developing breast cancer). You obtain records for thousands of women who died of breast cancer and contact any surviving daughters. Of those daughters who agree to participate, you randomly select 100 for genetic testing. X = the number out of 100 who test positive for the genetic marker.

   _____

Name _____ Date _____

2. According to the U.S. Centers for Disease Control and Prevention, 20% of the U.S. population is infected with the herpes simplex-2 virus. Suppose a person had three sex partners in the last year. Develop a Binomial probability distribution for the random variable X = number of partners out of three who were infected with this virus.

   a. Use the event-tree approach to list all possible outcomes of having three partners who were either infected (I) or not infected (N) with this virus, and write the value of X corresponding to these outcomes for each.

Event Tree	Outcome	X	P[Outcome]
	_____	_____	_____
	_____	_____	_____
	_____	_____	_____
	_____	_____	_____
	_____	_____	_____
	_____	_____	_____
	_____	_____	_____
	_____	_____	_____

   b. Use the multiplication rule to determine the probability for each of these outcomes. Write these probabilities to the right of the outcomes listed in the event tree. *Note:* If 20% of the population is infected, the probability that any randomly selected individual would be infected is $P = 0.2$.

   c. For each value X = number of infected partners out of three, write the number of different outcomes (combinations) that result in that value of X. Write the number in the first row of the table below.

Sample Space of X:	0	1	2	3
Number of Combinations	_____	_____	_____	_____
$P[X]$	_____	_____	_____	_____

   d. Use the addition rule to determine the probability associated with each value of X, based on the probabilities of outcomes and the number of combinations that result in each value of X. Enter these probability values in the second row of the table in part (c).

   e. Determine the probability that a person who has had three sex partners in the last year was exposed to the herpes simplex-2 virus *at least once* (i.e., $P[X \geq 1]$). List all possible values of X that meet the criterion described above. Use the rules of probability to determine $P[X \geq 1]$ described above.

3. It is well known that the risk of being exposed to a sexually transmitted disease increases as the cumulative number of sex partners increases. Use the Binomial table to determine the probability that a person would be exposed to the herpes simplex-2 virus *at least once* as a function of the cumulative number of sex partners. Enter the probabilities in the table below. For this exercise P = 0.2 (20% of the population are infected with this virus) and n = the number of partners.

n	1	2	4	8	16
$P[X \geq 1]$	_____	_____	_____	_____	_____

4. Use the Binomial formula to compute the probability that a person would be exposed to the herpes simplex-2 virus *at least once* if that person had 32 sex partners during his or her life. *Note:* $P[X \geq 1] = 1 - P[X = 0]$.

   Given P = 0.2 and n = 32; $P[X \geq 1]$ = _____. Show your calculations.

5. Suppose X is a Binomial random variable with n = 10 and P = 0.17 and Y is a Binomial random variable with n = 14 and P = 0.37.

   a. Use MS Excel to compute the probabilities associated with all values of X and Y. In column A, type the sample space for X beginning with 0 in cell A1. Click on the empty **cell B1**, then click on the = symbol in front of the long blank white area located just above the column letters. In this cell, type **BINOMDIST(A1,10,0.17,0)**. By first clicking on the = symbol, you instruct Excel to treat the contents of cell B1 as a formula rather than just text. **A1** in the formula indicates you want a Binomial probability for the X = 0 value in cell A1. Copy this formula down the B column to compute the probability for each value of X in the sample space. Repeat this procedure for random variable Y in columns C and D. Print out the part of the spreadsheet that contains these probability distributions and append to this page.

   b. Compute the probability that X will be an odd number.

   c. Compute the probability that Y will be an odd number.

   d. Compute the probability that X and Y will be odd numbers.

6. Given the values for n and P in parts (a) through (d), use the Binomial table to determine the probability of the events listed.

n	*P*	Event		*P*[Event]
a. $n = 8$	$P = 0.03$	$P[X \leq 2]$	$=$	_____
		$P[X < 2]$	$=$	_____
b. $n = 15$	$P = 0.4$	$P[X \geq 10]$	$=$	_____
		$P[X > 10]$	$=$	_____
c. $n = 7$	$P = 0.10$	$P[X = 0]$	$=$	_____
		$P[X < 2]$	$=$	_____
d. $n = 10$	$P = 0.25$	$P[X \geq 5]$	$=$	_____
		$P[X > 0]$	$=$	_____

7. Given that X is a Binomial variable with $n = 22$ and $P = 0.75$, use the Binomial formula to determine the probability $X > 20$.

8. Suppose that both parents in a family carry genes for blood types A and B (i.e., they are both type AB). The blood types of their children are independent and each child has a probability of 0.25 of having blood type A. There are four children in the family. Let X be the number of children out of $n = 4$ who have blood type A.

a. Use the Binomial table to determine the probability distribution of X and write this distribution in the format given below.

X		
P[X]		

b. Using the probability distribution in part (a), what is the probability that all four children will have blood type A?

c. What is the probability that none of the children will have blood type A?

Name _____ Date _____

Supplemental Problems

9. Do the following random variables meet the criteria to be Binomial variables? Explain.

  a. Natural red hair is an unusual trait in the U.S. population. It would be difficult to determine the proportion of the U.S. population with this trait because it is quite likely that a sample would not include *any* red-haired individuals. You decide to continue sampling until you have 10 red-haired individuals in the sample. X = the number of red-haired individuals in the sample and n = the total number of individuals sampled.

  b. Tay Sachs disease is a fatal genetic disorder that usually occurs in children of Jewish parents whose families immigrated from eastern Europe. You want to determine the proportion of people who carry the gene for this disorder in an isolated population of 250 Jews descended from Polish immigrants to the U.S. who established a communal farm in rural Pennsylvania. You perform blood tests for the presence of the gene on n = 20 individuals. X = the number who carry the Tay Sachs gene.

10. A couple plans to have four children. If we consider only the gender of the children, having four children is much like flipping a coin four times. There are 16 possible outcomes, and all can be assumed to be equally likely.

  a. Let the random variable X = the number of girls among four children. Use the event-tree approach to list all 16 possible outcomes of having four children and write the value of X corresponding to these outcomes for each.

Event Tree	Outcome	X	P[Outcome]

b. Use the multiplication rule to determine the probability for each of these outcomes, and write these probabilities to the right of the outcomes listed in the event tree.

c. For each value X = number of girls out of four births, write the number of different outcomes (combinations) that result in that value of X. Write in the first row of the table below.

Sample Space of X:	0	1	2	3	4
Number of Combinations	_____	_____	_____	_____	_____
$P[X]$	_____	_____	_____	_____	_____

d. Use the addition rule to determine the probability associated with each value of X, based on the probabilities of outcomes and the number of combinations that result in each value of X. Enter these probability values in the second row of the table above.

e. Determine probabilities for the following events. Show your work by listing all values of X that meet the specified criterion; then sum the probabilities.

The probability that the couple will have at least one girl _____

The probability that the couple will have two girls _____

The probability that the first child is a girl _____

11. Given the values for sample size n and assumed population proportion P below, use the Binomial table to determine the probabilities for the events listed. Show your work.

	n	P	Event	Probability	Work
a.	5	0.05	$P[X < 2]$	_____	_____
b.	6	0.25	$P[X \geq 4]$	_____	_____
c.	15	0.01	$P[X = 0]$	_____	_____
d.	18	0.40	$P[3 < X \leq 6]$	_____	_____
e.	12	0.10	$P[0 \leq X < 3]$	_____	_____

12. According to a CDC report, 29% of the U.S. population is infected with one of several incurable viral sexually transmitted diseases (herpes, human papillomavirus, hepatitis B, and HIV). Use the Binomial formula to compute the probability that a person who had five sex partners would be exposed to an incurable viral STD *at least once*.

Study Problems for Chapter 4 (with answer key)

1. Given the genetics of a couple, there is a probability of 0.25 that a child born to them will have a genetic disease. If these parents want a family of three children, develop the probability distribution for X = number of children that will have the disease:

   a. Use the event tree approach to list all possible outcomes of having three children in the space below (b).

   b. Determine the probabilities of each outcome and write the probability after each outcome in the event tree below.

   c. Combine the outcomes that produce the same value for X to obtain probabilities for each value of X in the sample space.

Sample Space of X:	0	1	2	3	4
$P[X]$	____	____	____	____	____

2. Using the probability distribution derived for Problem 1:

   a. What is the probability that this couple will have at least one child out of three with the disease?

   b. What is the probability that none of their three children will have the disease?

   c. What is the probability that the first child and the second child born to this couple will both be normal? Assess the validity of any assumptions needed to perform this calculation.

3. Given that X is a Binomial random variable, with $n = 20$ and $P = 0.4$, use the Binomial table to determine the following probabilities:

    a. $P[X \geq 10]$ = _____   **b.** $P[4 < X \leq 8]$ = _____

    c. $P[X > 1]$ = _____   **d.** $P[X = 5]$ = _____

4. A student is doing an experiment to determine whether or not salamanders are attracted to their own scent marking. He places each of 10 salamanders in a box that has three exit tubes. Filter paper containing each salamander's own scent is randomly placed in one of the 3 exit tubes. Even if the salamanders cannot smell their own scent, they will move into the arm of the maze that contains their scent one out of three times by random chance ($P = 0.33$ if salamanders are not attracted to their own scent). The student observes that 8 of the 10 salamanders moved into the arm of the maze that contained their own scent marking. He wants to determine the probability $P[X \geq 8$ if $n = 10$ and $P = 0.33]$. Use the Binomial formula to determine this probability. Show your work.

Answer Key for Chapter 4 Study Problems

1. **a., b.** N = normal child, D = child with genetic disease, X = number of diseased children

Outcome	X	P[Outcome]
NNN	0	$(0.75)(0.75)(0.75) = \mathbf{0.4219}$
NND	1	$(0.75)(0.75)(0.25) = \mathbf{0.1406}$
NDN	1	$(0.75)(0.25)(0.75) = \mathbf{0.1406}$
NDD	2	$(0.75)(0.25)(0.25) = \mathbf{0.0469}$
DNN	1	$(0.25)(0.75)(0.75) = \mathbf{0.1406}$
DND	2	$(0.25)(0.75)(0.25) = \mathbf{0.0469}$
DDN	2	$(0.25)(0.25)(0.75) = \mathbf{0.0469}$
DDD	3	$(0.25)(0.25)(0.25) = \mathbf{0.0156}$

c.

Sample Space of X:	0	1	2	3
P[X]	0.4219	0.4218	0.1407	0.0156

2. **a.** $P[\text{At least one child with disease}] = P[X \geq 1] = 1 - P[X = 0]$

$$= 1 - 0.4219 = \mathbf{0.5781}$$

b. $P[\text{No children will have the disease}] = P[X = 0] = \mathbf{0.4219}$

c. $P[\text{First child normal and second child normal}] = P[\text{First child normal}] \times P[\text{Second child normal}] = 0.75 \times 0.75 = \mathbf{0.5625}$

Whether or not one child is born with the disease in no way influences whether or not the next child will be born with the disease. Hence, these two events are independent.

3. **a.** $P[X \geq 10] = P[X = 10, 11, 12, ..., \text{or } 20]$

$$= 0.1171 + 0.0710 + 0.0355 + 0.0146 + 0.0049 + 0.0013 + 0.0003$$

$$= \mathbf{0.2447}$$

b. $P[4 < X \le 8] = P[X = 5, 6, 7, \text{ or } 8]$

$$= 0.0746 + 0.1244 + 0.1659 + 0.1797 = \mathbf{0.5446}$$

c. $P[X > 1] = 1 - P[X = 0 \text{ or } X = 1] = 1 - [0.0000 + 0.0005] = \mathbf{0.9995}$

d. $P[X = 5] = \mathbf{0.0746}$

4. $\dfrac{10!}{8! \, (10 - 8)!} \, (0.33)^8 \, (0.67)^{10-8} = 45 \, (0.0001406) \, (0.4489) = \mathbf{0.0028401}$

$\dfrac{10!}{9! \, (10 - 9)!} \, (0.33)^9 \, (0.67)^{10-9} = 10 \, (0.00004641) \, (0.67) \quad = \mathbf{0.0003109}$

$\dfrac{10!}{10! \, (10 - 10)!} \, (0.33)^{10} \, (0.67)^{10-10} = 1 \, (0.0000153) \, (1) \quad = \mathbf{0.0000153}$

$$\text{Answer} = \mathbf{0.0031663}$$

Exercise A: Simulating a Binomial Random Phenomena

Objectives You will be able to use a random numbers generator to simulate simple random phenomena (Bernoulli trials). Through direct observation of many outcomes for a Binomial random variable, you will be able to describe the nature of a random phenomenon.

Introduction One definition of the *probability of an event* is that it is the long-term relative frequency of the occurrence of that event. That is, if you observe a very large number of outcomes of a random phenomenon, the probability associated with any one specific type of outcome (event) is the proportion of all outcomes when that particular event occurred. For example, if you wanted to determine the probability that a coin tossed will land with the "heads" side on top, you could toss that coin thousands of times and determine the proportion (relative frequency) of tosses when the coin came up heads. There is a story that a student of statistics who fought in World War I was captured and imprisoned for many years. To pass the time, he tossed a coin tens of thousands of times to determine the precise probability of getting a heads. However, few people have time to spend tossing a coin thousands of times, and the probability derived for one coin may not be applicable to another coin. There is also the issue of precision in the estimate one makes. As you would likely guess, the greater the number of observations of the random phenomenon, the more precise the estimate of probability will be. So how can we use the concept of probability as long-term relative frequency?

    With computers, it is now possible to *simulate* random phenomena and study the results of thousands, millions, or even billions of outcomes. The computer user stipulates rules that define the nature of the random phenomenon, and the computer program quickly generates numerous outcomes from a random phenomenon that behaves according to those rules. Through simulation, we are able to better describe the "expected" probabilities (long-term relative frequencies) associated with particular outcomes or events from a random phenomenon that behaves according to the stipulated rules. If empirical probabilities derived by observation of the random phenomenon in the real world deviate from these simulated probabilities, this could be evidence that the real-world rules for the random phenomenon are not what you thought they were. For example, you assume that the probability of a single coin toss coming up heads is 0.5. Hence, it you toss a coin thousands of times, the relative frequency of "heads" should be 0.5. However, if the coin is not fair (that is, if it is unbalanced), such that one side is more likely than the other to land face up, the empirical relative frequency will not match the expected relative frequency derived from the simulation.

    Because computer simulations of random phenomena efficiently generate numerous outcomes, it is easy to observe what truly random phenomena look like. Most people have an incorrect conception of what it means to be random. For example, if a coin is tossed many times, most people expect heads and tails outcomes to be evenly dispersed. In fact, outcomes of truly random phenomena are often "lumpy," with occasional long sequences of identical outcomes (called "runs"). Computer simulations rapidly provide a sufficiently large number of outcomes to demonstrate these runs in outcomes of random phenomena.

Simulating Tosses of a Fair Coin

If I handed you a coin and asked you to determine whether or not it was fair, you would probably toss the coin some number of times to determine if $P[\text{Heads}] = P[\text{Tails}] = 0.5$. For example, you might toss the coin 10 times, expecting a fair coin to come up heads 5 of 10 times. However, it would probably not be surprising if the outcome of this random experiment was 4 heads or 6 heads. However, if the outcome was 1 head or 9 heads, you would likely conclude that this is impossible for a fair coin. Hence, the coin must be biased. However, this conclusion is not necessarily correct. That is, it is possible to get 9 heads out of 10 tosses of a truly *fair* coin, but unlikely. In this exercise you will simulate performing 500 experiments of tossing a coin 10 times and determining the relative frequency of values for the Binomial random variable $X =$ the number of heads out of $n = 10$ coin tosses.

Rules for a Fair Coin

1. There are only two outcomes, heads or tails.

2. These two outcomes are equally likely, $P[\text{Heads}] = P[\text{Tails}] = 0.5$.

Note: A random variable that has only two possible outcomes, with fixed probabilities associated with each outcome, is called a *Bernoulli variable*. Typically, when a computer simulates a Bernoulli variable, the two possible outcomes are coded as 0 and 1. For example, to simulate tossing a coin to determine the probability of getting "heads," you would stipulate that 1 = heads and 0 = tails. You always assign the value 1 to the outcome of interest.

Simulating a Bernoulli Random Variable

Many statistics computer programs provide options for generating random numbers for a variety of random phenomena. See Tutorial 4 for directions specific to the software used in your class. The directions below are generic instructions for what you must do.

1. Stipulate that you want to simulate **n = 10** Bernoulli observations, with **probability = 0.5**.

2. Stipulate that you want to perform this simulation **500** times. This will generate 500 variables, each with 10 values of either 0 or 1.

3. For each of the 500 variables that contain $n = 10$ Bernoulli observations, compute the summary statistic **Sum** $(X) =$ the count of "heads," coded as 1.

4. The sample space for $X =$ the number of heads out of $n = 10$ coin tosses is {0, 1, 2, 3, 4, 5, 6, 7, 8, 9, 10}. The purpose of this simulation is to determine the likelihood of getting each of these possible outcomes. You will determine this empirically using the frequency and relative frequency for each possible outcome out of the 500 simulated repetitions of this random phenomenon. Your statistics software should be able to quickly summarize the results of the 500 repetitions.

    a. Create a frequency table for values of the Sum variable computed using the 500 simulated repetitions of flipping a fair coin 10 times. This table will list how many times out of 500 each of the values in the sample space occurred.

    b. Next to these frequencies for each value of the Sum, compute the relative frequency ($= X/500$).

Report   Put the frequency table for the Sum variable X on a single word-processor page with the title **Simulation of 500 Experiments of Tossing a Coin 10 Times**. Write the relative frequency values next to each of their corresponding frequency values in the table. Based on these results, *type answers to the following questions*, in complete sentences, on that page.

1. If the coin is "fair," describe the "expected" outcome of flipping the coin 10 times.

2. Based on the results of your simulation, what is the relative frequency (empirical probability) of this expected outcome?

3. Suppose you were a sheriff in a wild west town. A shady looking gambler has been winning money from the townsfolk on a game of chance based on flipping a coin and gambling on whether or not it comes up heads. The gambler always uses his own coin, always bets it will come up heads, and has won much more than he has lost. The angry townsfolk believe he is a cheat and want the sheriff to run him out of town on a rail. As sheriff, you want more proof of cheating than the grumbling of sore losers. You flip the gambler's coin 10 times to see if it is fair (heads and tails equally likely).

   Based on the results of your simulation, what outcome from this test based on 10 coin tosses would give you (the sheriff) enough evidence to convict the gambler of cheating and so warrant giving the gambler "the treatment"? Explain the reasoning behind the criterion you choose, based on the relative frequencies from your simulation. Your justification should address the following points:

   a. Relative frequencies computed from this simulation are estimates of the probabilities associated with the possible outcomes of flipping a *fair* coin 10 times.

   b. Hence, the probability associated with your criterion for convicting the gambler is the probability that you would convict an innocent person.

   c. Our legal system and process for doing science try to minimize the probability that we would make an erroneous decision, but the only way to completely avoid error is to never make a decision. This leads to paralysis and stagnation. Hence, you are *not* allowed to beg-off by complaining about inadequate data. You *must* make a decision based on the 10 coin tosses. You might consider trying to apply the beyond-a-reasonable-doubt criterion used in our criminal justice system. Scientists are always required to draw conclusions based on less data than they would like.

4. Go back to your random numbers generator and generate one Bernoulli variable with 500 observations (0s and 1s) and a probability of success = 0.5. This simulates tossing a fair coin 500 times. Scroll down the 500 values and determine the longest run of consecutive 0s and the longest run of consecutive 1s. While scrolling, pay attention to all sequential runs of the same value.

   Longest run of 0s (number of consecutive 0s) _____
   Longest run of 1s (number of consecutive 1s) _____

   Record these observations and describe how these are relevant to the statement "Outcomes of random phenomena are lumpy."

2. Use the Standard Normal Table to determine Z-values that correspond to the following probabilities.

$P[Z \geq$ _____ $] = 0.10$ $P[Z \leq$ _____ $] = 0.10$

$P[Z \leq$ _____ $] = 0.75$ $P[Z \geq$ _____ $] = 0.50$

$P[Z \geq$ _____ $] = 0.40$ $P[Z \leq$ _____ $] = 0.95$

3. Assume that in each case below the random variable X has a Normal distribution with a population mean (μ) and standard deviation (σ) as given. Use the Empirical Rule to determine approximate probabilities for the specified events. Then convert the X-values to Z-values and determine the exact probabilities using the Standard Normal Table.

μ	σ	Event	Approx. Prob.	Z-value	Exact Probability
100	20	$P[X \geq 120]$	_____	_____	_____
75	15	$P[X \leq 45]$	_____	_____	_____
20	3	$P[X \geq 29]$	_____	_____	_____
8	2	$P[4 \leq X \leq 10]$	_____	_____	_____

4. Blood triglyceride concentration is one of a number of variables used in assessing risk of coronary heart disease. Suppose that the distribution of this variable is approximately Normal, with a mean $\mu = 165$ and a standard deviation $\sigma = 55$.

a. Scale the X-axis of the accompanying Normal curve in accordance with this description. Shade-in the area under the curve associated with the proportion of the population that has blood triglyceride concentration 250 or more.

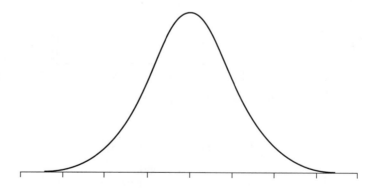

b. Use the Standard Normal Table to determine this proportion (= probability).

Answer Key for Chapter 5 Study Problems

1. $P[Z \leq -1.04]$ **0.1492** $P[0 \leq Z \leq 2.05]$ **0.4798** $(0.9798 - 0.5)$

   $P[Z \geq 0.71]$ **0.2389** $P[-1.90 \leq Z \leq 0]$ **0.4713** $(0.5 - 0.0287)$

   $P[Z \geq -0.93]$ **0.8238** $P[-2.11 \leq Z \leq 1.67]$ **0.9351** $(0.9525 - 0.0174)$

2. $P[Z \geq +\mathbf{1.285}] = 0.10$ $P[Z \leq -\mathbf{1.285}] = 0.10$

   $P[Z \leq +\mathbf{0.675}] = 0.75$ $P[Z \geq \mathbf{0}] = 0.50$

   $P[Z \geq +\mathbf{0.255}] = 0.40$ $P[Z \leq \mathbf{1.645}] = 0.95$

3.

μ	σ	Event	Approx. Prob.	Z-value	Exact Probability
100	20	$P[X \geq 120]$	**0.16**	**+1.00**	**0.1587**
75	15	$P[X \leq 45]$	**0.025**	**-2.00**	**0.0228**
20	3	$P[X \geq 29]$	**0.005**	**+3.00**	**0.0013**
8	2	$P[4 \leq X \leq 10]$	**0.815**	**$-2 \leq Z \leq +1$**	**0.8185**

4. a.

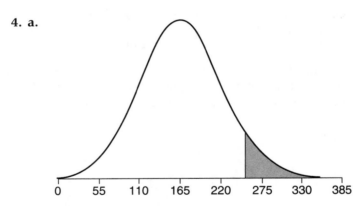

   b. $Z = (250 - 165) / 55$

   $= +1.55$

   $P[X \geq 250] = P[Z \geq 1.55]$

   $= 0.0606$

Name _____   Date _____

8. A doctor assesses a patient for hypokalemia (low blood potassium). Variation of the actual potassium level in the patient's blood and measurement error for the blood test both cause repeated measurements of the patient's blood potassium level to vary according to a Normal distribution with mean $\mu = 3.8$ and standard deviation $\sigma = 0.2$. The patient would be diagnosed as hypokalemic if blood tests show a potassium concentration of 3.5 or less.

   a. If potassium concentration for this *healthy* patient was estimated based on two blood tests ($n = 2$), what is the probability that she will be diagnosed as hypokalemic? That is, what is $P[\bar{x} \le 3.5]$ even though $\mu = 3.8$?

   Describe the sampling distribution of the mean.

   Center: $E(\bar{x})$ = _____

   Spread: $\sigma_{\bar{x}}$ = _____

   Shape: _____

   Scale the X-axis of the sampling distribution and shade the area under the curve that corresponds to the probability that this healthy patient would be diagnosed as having hypokalemia.

   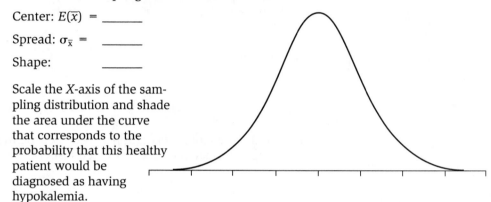

   Based on this sampling distribution, compute the probability of the event that the mean blood potassium level for two blood tests will be 3.5 or less (blood test indicates hypokalemia), even though the patient is healthy ($\mu = 3.8$).

b. If a blood sample was taken on four separate days, and the mean potassium concentration for these $n = 4$ samples was compared to the 3.5 criterion, what is the probability that this healthy patient would be diagnosed as hypokalemic?

Describe the sampling distribution of $\bar{x}$.

Center: $E(\bar{x})$ = _____

Spread: $\sigma_{\bar{x}}$ = _____

Scale the *X*-axis and shade the area under the curve that corresponds to the probability that this healthy patient would be diagnosed as having hypokalemia.

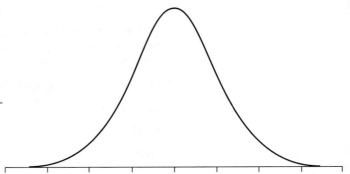

Compute the probability $P[\bar{x} \le 3.5]$. (Tests indicate hypokalemia for this patient if $\bar{x}$ was based on a sample of $n = 4$ blood tests.)

c. If we didn't know the distribution of daily blood potassium measurements was Normal, would this influence what you did for part (b)? Explain.

d. What can be done to ensure that the sampling distribution of a sample mean *is* Normal?

9. On what basis can we make the assumption that the center of the sampling distribution of a statistic is the true value of the population parameter? (*Hint:* It has something to do with study design concepts presented in Chapter 1.)

10. A general rule in study design is: "The larger the sample size, the better the study." Explain why this statement is true based on characteristics of sampling distributions as described by:

   a. The Law of Large Numbers.

   b. The Central Limit Theorem.

11. In your own words, explain the concept of the sampling distribution of $\bar{x}$.

Supplemental Problems

12. A student is doing an experiment to determine whether or not salamanders are attracted to their own scent marking. He places each of 50 salamanders in a box that has three exit tubes. Filter paper containing each salamander's own scent is randomly placed in one of the three exit tubes. Even if the salamanders cannot smell their own scent, they will move into the arm of the maze that contains their scent one out of three times by random chance (**P** = 0.33 if salamanders are not attracted to their own scent). The student observes that X = 25 out of 50 ($\hat{p}$ = 0.5) salamanders moved into the arm of the maze that contained their own scent marking. He wants to determine the probability $P[\hat{p} \geq 0.5$ if n = 50 and **P** = 0.33].

a. Describe the sampling distribution of $\hat{p}$—including center, spread, and shape—under the assumption that salamanders are *not* attracted to their own scent marking.

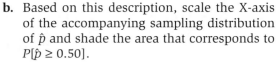

Center: $E(\hat{p})$ = _____

Spread: $\sigma_{\hat{p}}$ = _____

Shape: _____

b. Based on this description, scale the X-axis of the accompanying sampling distribution of $\hat{p}$ and shade the area that corresponds to $P[\hat{p} \geq 0.50]$.

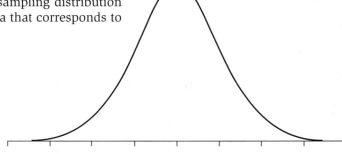

c. Use this sampling distribution and the Standard Normal Table to determine the probability $P[\hat{p} \geq 0.5]$.

Name _____   Date _____

13. Suppose that the body mass of an insect species is a Normally distributed variable
with mean μ = 11 mg and standard deviation σ = 2.6 mg. For each of the following,
scale the *X*-axis of the accompanying sampling distribution curve, shade the area under
the curve that corresponds to the probability associated with the specified value range,
and determine this probability.

a. What is the probability
that a single insect
(*n* = 1) from this pop-
ulation would have a
body mass 10 mg or
less?

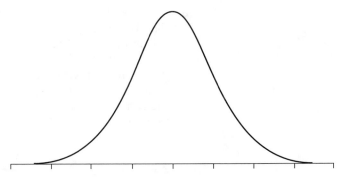

b. What is the probability
that the mean mass for
a sample of *n* = 10
insects from this popu-
lation would be 10 mg
or less?

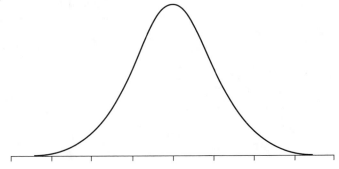

c. What is the probability
that the mean mass for
a sample of *n* = 20 in-
sects from this popula-
tion would be 10 mg
or less?

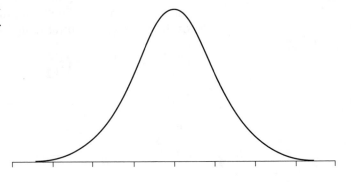

14. A national survey indicates that 10% of people in the age class 18 to 24 are infected with the genital herpes simplex-2 virus. You want to know if the proportion of students at your university who have this sexually transmitted disease is less than this national value. Suppose that you were able to obtain blood test results for a randomized sample of $n = 300$ students, of which $X = 28$ tested positive for the genital herpes virus.

a. Describe the sampling distribution of $\hat{p}$ under the assumption that the proportion of students infected with the genital herpes virus is the same as the national value.

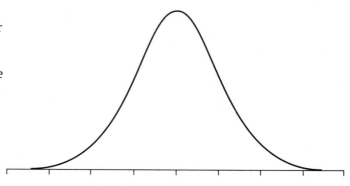

Center: _____

Spread: _____

Shape: _____

b. Scale the X-axis appropriately, and shade the area under the curve that corresponds to $P[\hat{p} \leq 0.0933$ if $n = 300$ and $P = 0.10]$. Use this sampling distribution to determine the probability $P[\hat{p} \leq 0.0933]$.

15. Suppose that the population distribution for body mass measurements on adult humans is positively skewed and bimodal (due to gender differences), with $\mu = 70$ kg and $\sigma = 20$ kg. For each of the following sample sizes, describe the sampling distribution of the sample mean $\bar{x}$, including the center $E(\bar{x})$, spread $\sigma_{\bar{x}}$, and shape. Apply the rules of thumb for the Central Limit Theorem.

	$n = 15$	$n = 40$	$n = 100$
$E(\bar{x})$:	_____	_____	_____
$\sigma_{\bar{x}}$:	_____	_____	_____
Shape:	_____	_____	_____

16. Suppose that the wing span of a rare and endangered bat species is a Normally distributed variable with $\mu = 30$ cm, with $\sigma = 2.5$ cm. A previously unrecorded population of this species is found, and $n = 10$ bats are captured and measured for many variables, including wing span.

   a. Determine the probability that this sample of 10 bats would have an average wing span $\bar{x} \leq 29$ cm if the newly discovered population had the same distribution of wing spans as that described for other populations. First, scale the X-axis of the sampling distribution and shade the area that corresponds to $P[\bar{x} \leq 29]$. Then use the Standard Normal Distribution to determine the exact probability.

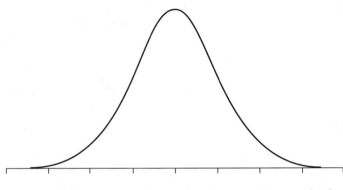

   b. What would be the probability $P[\bar{x} \leq 29]$ if this sample mean from the new population was based on a sample size $n = 20$? Rescale the X-axis of the sampling distribution of the mean and shade the area that corresponds to $P[\bar{x} \leq 29]$. Use the Standard Normal Distribution to determine the probability.

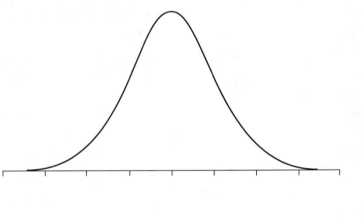

   c. Explain why the probability $P[\bar{x} \leq 29]$ differed between the analysis based on $n = 10$ and the analysis based on $n = 20$.

Name _____   Date _____

Study Problems for Chapter 6

1. Given the hypothetical population proportion value **P** and the sample size n, describe the sampling distribution of the sample proportion $\hat{p}$. If it is Normal, use the Standard Normal Table to determine the probability of obtaining a value for $\hat{p}$ in the specified range. If the distribution is Binomial, use the Binomial table.

P	**n**	**E($\hat{p}$)**	$\sigma_{\hat{p}}$	**Shape (Binomial or Normal?)**	
0.7	40	_____	_____	_____	$P[\hat{p} \geq 0.8]$ = _____
0.4	80	_____	_____	_____	$P[\hat{p} > 0.5]$ = _____
0.2	20	_____	_____	_____	$P[\hat{p} \leq 0.05]$ = _____
0.9	150	_____	_____	_____	$P[\hat{p} \leq 0.85]$ = _____

2. Given the hypothetical population mean (μ) and standard deviation (σ) for a measured variable X and the sample size, describe the sampling distribution of the mean $\bar{x}$ if the shape of the data distribution is approximately Normal. Use this sampling distribution to determine the probability of obtaining a value for $\bar{x}$ in the specified range.

μ	σ	**n**	**E($\bar{x}$)**	$\sigma_{\bar{x}}$	
25	6	20	_____	_____	$P[\bar{x} \leq 21]$ = _____
80	13	50	_____	_____	$P[\bar{x} \geq 85]$ = _____
230	41	60	_____	_____	$P[\bar{x} \geq 240]$ = _____
115	35	75	_____	_____	$P[\bar{x} \leq 100]$ = _____

Name _____   Date _____

3. Based on the boxplots and the Normal quantile plots below, use the "rules of thumb"
for applying the Central Limit Theorem to determine the minimum sample size that
would be required before you would assume that the sampling distribution of the mean
is Normal. Explain your answer.

   a. Minimum sample size: _____

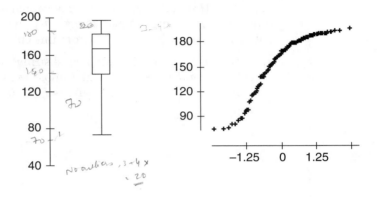

   Explanation:

   b. Minimum sample size: ____ 40

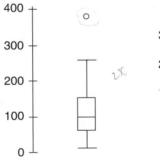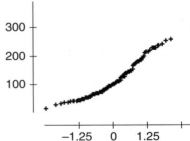

   Explanation:

4. A student uses controlled breeding to produce a colony of fruit flies in which all individuals should be heterozygous for a wing length trait. (That is, all should carry one gene for long wings and one gene for short wings, and the flies all should have long wings.) While transferring flies to new culture bottles, she thinks that perhaps her pure colony may have been contaminated by a few stray flies that are homozygous "long wing." (These flies have long wings, but have *two* long-wing genes.) If the colony is pure, 25% (**P** = 0.25) of the offspring should exhibit the recessive trait (short wings). If the colony is contaminated, fewer offspring will have short wings. Suppose that 36 out of 180 offspring ($\hat{p}$ = 0.20) from this culture exhibit the short-wing trait.

a. Scale the X-axis of the sampling distribution for $\hat{p}$ under the assumption that the colony is pure (**P** = 0.25), with sample size n = 180. Shade the area under the curve that corresponds to the probability $P[\hat{p} \leq 0.20$ if **P** = 0.25].

$E(\hat{p})$ = _____

$\sigma_{\hat{p}}$ = _____

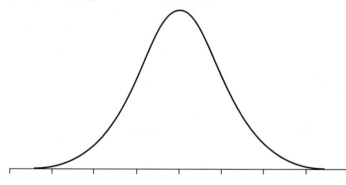

b. Use the Standard Normal Table to determine the probability $P[\hat{p} \leq 0.20$ if **P** = 0.25].

5. The taxonomic description for the gray wolf, based on data from the lower 48 states of the United States, says that the mean mass of an adult male is $\mu = 38$ kg, with a standard deviation $\sigma = 10$. A biologist studying gray wolves in northern Alaska obtains data for $n = 25$ adult male wolves, with mean adult male mass $\bar{x} = 41$ kg. Bergmann's rule says that mean body mass for individuals of a single mammal species is greater in cold environments than in warm environments. The biologist wants to know the probability of obtaining a sample mean $\bar{x} \geq 41$ if the adult male wolves in the Alaskan population have the same body mass distribution as wolves in the lower 48 states.

a. Scale the X-axis of the sampling distribution for $\bar{x}$ if $\mu = 38$, $\sigma = 10$, and $n = 25$. Shade the area that corresponds to the probability $P[\bar{x} \geq 41$ if $\mu = 38]$.

$E(\bar{x}) =$ _____

$\sigma_{\bar{x}} =$ _____

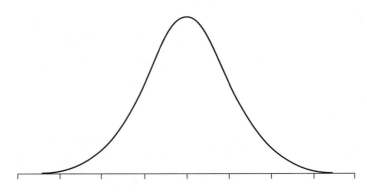

b. Use the Standard Normal Table to determine the probability $P[\bar{x} \geq 41$ if $\mu = 38]$.

Name _____   Date _____

Answer Key for Chapter 6 Study Problems

1.

P	n	$E(\hat{p})$	$\sigma_{\hat{p}}$	Shape		
0.7	40	0.7	0.0725	Normal	$P[\hat{p} \geq 0.8] = \mathbf{0.0838}$	$(Z \geq +1.38)$
0.4	80	0.4	0.0548	Normal	$P[\hat{p} \leq 0.25] = \mathbf{0.0031}$	$(Z \leq -2.74)$
0.2	20	0.2	0.0894	Binomial	$P[\hat{p} \leq 0.05] = \mathbf{0.0691}$	$(X = 0 \text{ or } 1)$
0.9	150	0.9	0.0245	Normal	$P[\hat{p} \leq 0.85] = \mathbf{0.0207}$	$(Z \leq -2.04)$

2.

μ	σ	n	$E(\bar{x})$	$\sigma_{\bar{x}}$			
25	6	20	25	1.34	$P[\bar{x} \leq 21]$	$= \mathbf{0.0014}$	$(Z \leq -2.99)$
80	13	50	80	1.84	$P[\bar{x} \geq 85]$	$= \mathbf{0.0033}$	$(Z \geq +2.72)$
230	41	60	230	5.29	$P[\bar{x} \geq 240]$	$= \mathbf{0.0294}$	$(Z \geq 1.89)$
115	35	75	115	4.04	$P[\bar{x} \leq 100]$	$= \mathbf{< 0.0001}$	$(Z \leq -3.71)$

3. a. Minimum sample size = **15**. The data distribution is skewed, but there are no outliers.

 b. Minimum sample size = **40**. Data distribution is skewed *and* outliers are present.

4. a. $E(\hat{p}) = \mathbf{0.25}$

 $\sigma_{\hat{p}} = \mathbf{0.0323}$

 Shape: **Normal**

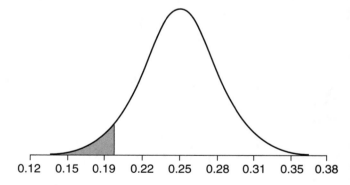

 b. $P[\hat{p} \leq 0.20 \text{ if } \mathbf{P} = 0.25, n = 180] = P[Z \leq (0.20 - 0.25) / 0.0323]$

 $= P[Z \leq -1.55]$

 $= \mathbf{0.0606}$

5. a. $E(\overline{x}) = 38$

$\sigma_{\overline{x}} = 2 = 10 / \sqrt{25}$

Shape: **Normal**

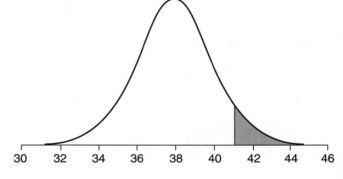

b. $P[\overline{x} \geq 41 \text{ if } \mu = 38] = P[Z \geq (41 - 38) / 2]$

$= P[Z \geq 1.50]$

$= \mathbf{0.0668}$

Exercise A: Simulation of the Sampling Distribution of the Mean

Objective   You will learn how the nature of the population and the sample size influence the characteristics (center, spread, shape) of the sampling distribution of the mean. You will also learn how changing the sample size influences the quality of estimates of the population mean, including its effects on precision and accuracy. Finally, you will learn how simulations of repeated sampling from a population can be used to obtain empirical sampling distributions of the mean that demonstrate the theoretical characteristics of such distributions.

Introduction   Although most scientific studies are interested in the characteristics of an entire population, information (data) is usually gathered for only a very small portion of the total population, called a *sample*. It is not feasible to measure entire populations. Because data collected on a sample will be used to describe the entire population, it is critical that the sample be representative of the larger population (*unbiased*), and that the estimates of population characteristics be repeatable (*precise*). Unbiased samples provide estimates that are equally likely to be greater than or less than the actual value for the population characteristic. The simplest procedure for obtaining unbiased estimates is to randomly sample individuals from the population. Precise estimates are those that give similar values if repeated samples are taken from the same population. Precise estimates are obtained by taking samples that include a large number of individuals. If a sample mean is computed using data for many individuals, unusual (*outlier*) values for a few individuals will have minimal influence on value of the sample mean. However, if only a few individuals are measured, one unusual value can substantially alter the values of the sample mean.

We must recognize that the estimates of population mean computed from samples exhibit random variation. Even if repeated samples are taken from exactly the same population, different samples will contain different individuals, and produce different estimates of population mean. The consequence of this *random sampling variation* is that we must treat sample statistics (our estimates of population parameters) as random variables.

There are two approaches for understanding the pattern of variation in values of a random variable, *theoretical* (based on rules of probability and logic) and *empirical* (based on many observations of the random variable in action). Mathematical statisticians have used the theoretical approach to derive descriptions of the probability distributions for sample statistics that exhibit random sampling variation. In this exercise, we will use a simulation of the empirical approach to describe the probability distribution for the sample mean. Based on relative frequencies of values for sample means derived from repeated sampling from the same population, we will construct empirical *sampling distributions of the mean* (probability distribution for the value of the sample mean). These distributions will display the amount of variation in the values of the sample mean, and which values are more/less likely to occur. This empirical approach is similar to flipping a coin many times to determine the probabilities associated with the heads and tails outcomes. We will study the influence of sample size on the characteristics (center, spread, shape) of these empirical sampling distributions. We will also study how differences in the nature of population distribution (Normal versus skewed) influences the characteristics of the sampling distribution of the mean. Finally, we will compare the characteristics of empirically derived sampling distributions to those predicted from theoretical derivations of the same distributions.

Theoretical Descriptions of the Sampling Distribution of the Mean

The theoretical description of the sampling distribution of the mean is based on a number of logical premises, theorems, and laws. First, a fundamental premise is that samples obtained by a randomized, unbiased method should produce sample means that are equally likely to be greater than or less than the true population mean. Hence, the expected value for the sample mean $E(\overline{x})$, which is at the center of the sampling distribution, is the true population mean μ. Second, the Law of Large Numbers states that as the number of observations in a representative sample increases, the value of a sample statistic is more likely to be close to the true population parameter value. The precision of any sample mean depends on 2 factors: (1) the sample size and (2) the amount of variability present in the population, as quantified by the population standard deviation (σ). The precision of the sample mean is quantified by the Standard Deviation of the Mean ($\sigma_{\overline{x}}$), and is computed using equation [1] below. This equation is based on a theoretical derivation. You can see that as sample size n increases, $\sigma_{\overline{x}}$ decreases and precision increases.

$$\sigma_{\overline{x}} = \sigma / \sqrt{n} \qquad\qquad [1]$$

■ **EXAMPLE**   Suppose a random sample of 25 observations was obtained from a population. If the population standard deviation for this variable is σ = 16, the standard deviation of the mean is:

$$\sigma_{\overline{x}} = 16 / \sqrt{25} = 3.2 \qquad\qquad ■$$

This number quantifies the amount of variability we expect to see in the values of sample means computed from many repeated, independent samples of size $n = 25$ taken from this population. Finally, the Central Limit Theorem states that as the sample size increases, the shape of the sampling distribution of the mean becomes Normal. Taken together, the theoretical description of the sampling distribution of the mean is: (1) it is centered over the true population mean, (2) its spread is determined by the amount of variation in the population (σ) and the sample size (n), and (3) its shape is Normal, if sample size is large enough. We will use simulations of repeated sampling from populations with known characteristics to empirically demonstrate this theoretical description of the sampling distribution of the mean.

Terminology for Simulating the Sampling Distribution of the Mean

Suppose each student in a class of 50 obtained a random sample of 40 students from their university, measured each for height, and computed a sample mean to estimate the true mean height of students at that university. Assuming the students followed instructions, each of their sample means computed from 40 height measurements would be a valid estimate of the true mean height of the student body. However, the 50 independent estimates of mean height produced by the class would exhibit random sampling variation. If we combined the 40 individual data values obtained by each of the 50 students in the class, we would have a total of 2,000 height measurements. A histogram of these *individual* height data values would provide a picture of the variability of height in the university student body. Such a histogram is called a **Data Distribution**. This distribution of individual data values provides an approximate picture of the nature (center, spread, shape) of the **Population Distribution** of height that would be obtained if we sampled all individuals in the population. A histogram of 50 *mean* height values (each computed from 40 students), obtained from repeated, independent samples from the same population would display the variability among estimates of mean height and would be an empirical **Sampling Distribution of the Mean**.

This Exercise includes two options (using two different software products) for performing a simulation of re-sampling to empirically demonstrate the sampling distribution of the mean:

The **Quadrat Sampling Simulator** [Oakleaf Systems, PO Box 472, Decorah, IA, 52101, (563) 382-4320] allows you to actually watch repeated random sampling of a simulated population, listing the raw data and mean for each sample as it occurs. In this option, students must hand tally the frequencies for various values of the sample means (more laborious), but the students get a more "hands-on" feel for the concept of re-sampling the same population to demonstrate random sampling variation of the mean.

The **Random Numbers Simulators** in many statistics software programs allow the student to generate thousands of repeated samples of any sample size, taken from a variety of population distributions (discrete, continuous, Normal, positively skewed, negatively skewed, and others). In seconds, the software computes the mean for each of thousands of samples and tallies the frequencies of different values for the sample mean obtained from these repeated samples to produce an empirical sampling distribution. This is the most efficient simulation option, but the student may be less likely to understand what the simulation represents when the computer does all the work.

Tutorials for implementing these different approaches for simulating the sampling distribution of the mean are available at the web site *http://math.jbpub.com/leblanc/*. Given the different nature of the results from these two simulation options, separate descriptions of the assignment for this Exercise are provided for each option.

Assignment for Option 1: Quadrat Simulation

You should submit your tally histogram data sheets for the Random and Clumped populations. Your answers to the following questions should be typed and double-spaced. Your answers should be *written in well-constructed paragraphs* that address all the points raised in the subquestions under each main question. In answering these questions, keep the following points in mind: (1) High *precision* is indicated when the values of sample means computed from repeated samples taken from the same population are tightly clumped together; (2) High *accuracy* is indicated if the 50 sample means are close to the "true value." In this case, with true density = 500 organisms per hectare, and the plot area = 0.01 hectare, the "true value" for mean plot density = 5 organisms/plot. (3) In actual practice, only one sample of n plot is measured, and the theoretical standard deviation of the mean computed from that sample is the only measure of precision for the estimate of the mean.

Questions   1. Describe the *data distribution* for density values from individual plots sampled from the Random population (tally histogram on the left side of the data page). Your description should include the center, spread (use the minimum and maximum to define spread), and shape of the distribution.

Compare the center, spread, and shape of the Random population data distribution to the empirical sampling distributions of the mean for sample sizes of n = 3 and n = 15 plots from this population. How are each of the sampling distributions for the mean different from, or similar to, the Random population data distribution?

2. Describe the *data distribution* of data values for individual plots sampled from the Clumped population, including the distribution center, spread, and shape.

Compare the center, spread, and shape of the Clumped population data distribution to the empirical sampling distributions of the mean for sample sizes of $n = 3$ and $n = 15$ plots from this population. How are each of the sampling distributions for the mean different from, or similar to, the Clumped population data distribution?

3. Describe how the comparisons of data distributions and sampling distributions of the mean you made for #1 and #2 demonstrate the Central Limit Theorem (CLT).

   Describe how the descriptions of the sampling distributions of the mean you made for #1 and #2 demonstrate the Law of Large Numbers (LLN).

4. It is a universal rule in the practice of science that larger sample sizes are always better than smaller sample sizes. Based on your answers to questions 1 to 3 above, explain the advantages of large sample sizes.

5. Compare the spread of the sampling distributions of the mean for $n = 15$ between the Random (low variability) and Clumped (high variability) populations. How does the level of variation in the population affect the precision of sample means? If you wanted to attain the same precision for estimates of the mean for the Clumped population as you attained with samples of $n = 15$ from the Random population, how would you need to modify the sampling from the Clumped population?

   Perform a similar comparison between sampling distributions of the mean for Random and Clumped populations obtained with sample size $n = 3$.

6. Exploring the random variation of sample means based on simulations of repeated sampling from the same population is similar to exploring the probability of getting heads/tails outcomes from flipping a coin many times. Each independent repetition of sampling 3 or 15 plots from the same population is analogous to flipping a coin once. The Law of Large Numbers tells us that the more repetitions of the random process we observe, the better will be our understanding of the probability distribution. The values listed below are the mean and standard deviation values for 550 repeated sample means computed from samples of $n = 3$ and $n = 15$ plots each. These values are empirical estimates of the true mean (μ) and the true standard deviation of the mean ($\sigma_{\bar{x}}$) for the sampling distributions of the mean for sample sizes $n = 3$ and $n = 15$.

	Random ($n = 15$)	Random ($n = 3$)	Clumped ($n = 15$)	Clumped ($n = 3$)
Mean of 550 $\bar{x}$'s	5.02	5.19	5.19	5.39
S.D. of 550 $\bar{x}$'s	0.65	1.37	1.61	4.04

The true population mean and standard deviation for *individual* plot density values in the Random population are $\mu = 5$, $\sigma = 2.5$ and for the Clumped population, $\mu = 5$, $\sigma = 7$.

a. Given these population standard deviations (σ's), use formula [1] in this Exercise to compute values of the "Theoretical" standard deviation of the mean ($\sigma_{\bar{x}}$) for sampling distributions of the mean based on sample sizes $n = 15$ and $n = 3$ plots for both the Random and Clumped populations. Enter these values in a table like the one shown below.

	Random (n = 15)	Random (n = 3)	Clumped (n = 15)	Clumped (n = 3)
$\sigma_{\bar{x}} =$				

b. Compare these values for $\sigma_{\bar{x}}$ computed using the formula derived from a theoretical proof to the Empirical estimates of the standard deviation of the mean, computed as the standard deviation of 550 $\bar{x}$ values obtained from repeated samples, as described above. Are the empirical and theoretical values similar? Why would we *not* expect the empirical values to be exactly equal to the theoretical values?

c. How might this comparison of empirical vs. theoretical values for the standard deviation of the mean be different if it had been based on the mean and standard deviation for your 50 sample means from repeated samples, rather than 550 means? Explain how this outcome is related to the Law of Large Numbers.

Random Dispersion Population

Number of Organisms	Tally of Individual Plot Counts for 150 10 × 10 m Plots
0	| _____
1	| _____
2	| _____
3	| _____
4	| _____
5	| _____
6	| _____
7	| _____
8	| _____
9	| _____
10	| _____
11	| _____
12	| _____
13	| _____
14	| _____
15	| _____
16	| _____
17	| _____
18	| _____
19	| _____
20	| _____
21	| _____
22	| _____
23	| _____
24	| _____
25	| _____
26	| _____
27	| _____
28	| _____
29	| _____
30	| _____
31	| _____
32	| _____
33	| _____
34	| _____
35	| _____
36	| _____
37	| _____
38	| _____
39	| _____
40	| _____

Tally of Sample Means for 50 Samples of n = 15 10 × 10 m Plots

0	| _____
1	| _____
2	| _____
3	| _____
4	| _____
5	| _____
6	| _____
7	| _____
8	| _____
9	| _____
10	| _____
11	| _____
12	| _____
13	| _____
14	| _____
15	| _____
16	| _____
17	| _____
18	| _____

Tally of Sample Means for 50 Samples of n = 3 10 × 10 m Plots

0	| _____
1	| _____
2	| _____
3	| _____
4	| _____
5	| _____
6	| _____
7	| _____
8	| _____
9	| _____
10	| _____
11	| _____
12	| _____
13	| _____
14	| _____
15	| _____
16	| _____
17	| _____
18	| _____

Name _____   Date _____

Clumped Dispersion Population

Number of Organisms	Tally of Individual Plot Counts for 225 10 × 10 m Plots	Tally of Sample Means for 50 Samples of n = 15 10 × 10 m Plots
0	|_____	0 |_____
1	|_____	1 |_____
2	|_____	2 |_____
3	|_____	3 |_____
4	|_____	4 |_____
5	|_____	5 |_____
6	|_____	6 |_____
7	|_____	7 |_____
8	|_____	8 |_____
9	|_____	9 |_____
10	|_____	10 |_____
11	|_____	11 |_____
12	|_____	12 |_____
13	|_____	13 |_____
14	|_____	14 |_____
15	|_____	15 |_____
16	|_____	16 |_____
17	|_____	17 |_____
18	|_____	18 |_____
19	|_____	

Tally of Sample Means for 50 Samples of n = 3 10 × 10 m Plots

Number of Organisms	Tally of Individual Plot Counts for 225 10 × 10 m Plots	Tally of Sample Means
20	|_____	
21	|_____	
22	|_____	0 |_____
23	|_____	1 |_____
24	|_____	2 |_____
25	|_____	3 |_____
26	|_____	4 |_____
27	|_____	5 |_____
28	|_____	6 |_____
29	|_____	7 |_____
30	|_____	8 |_____
31	|_____	9 |_____
32	|_____	10 |_____
33	|_____	11 |_____
34	|_____	12 |_____
35	|_____	13 |_____
36	|_____	14 |_____
37	|_____	15 |_____
38	|_____	16 |_____
39	|_____	17 |_____
40	|_____	18 |_____

Assignment for Option 2: Simulation Based On Random Number Generator

After completing the tutorial for Option 2 of the Chapter 6 Exercise, you should have the following statistical print-outs pasted into pages in a word processor document:

1. Five pages of graphics (a histogram and Normal quantile plot on each page) generated from results of the simulation of sampling from a Non-Normal population. Each page should have a descriptive title for the contents of that page.

2. One page that contains descriptive statistics for the simulated sampling from the Non-Normal population. These statistics include empirical estimates of the mean and standard deviation for the population distribution and the sampling distributions of the mean for sample sizes 5, 15, 40, and 100.

3. Four pages of graphics generated from results of the simulation of sampling from the Normal population. Page contents and layout should be the same as described for the Non-Normal population simulation.

4. One page that contains descriptive statistics for the data distribution and the sampling distributions of the mean generated by the simulation of sampling from the Normal population with sample sizes of 5, 15, and 40.

You should *type* answers to the following questions, double-spaced, *written in well-constructed paragraphs*. Each paragraph should address all the points raised in the subquestions under each main question. In answering these questions, keep the following points in mind: (1) High *precision* is indicated when the values of sample means from repeated samples taken on the same population are tightly clumped together. (2) High *accuracy* is indicated if the values of sample means from repeated samples from the same population are close to the true value of the population mean. (3) In actual practice, only one sample of n observations is obtained, and the theoretical standard deviation of the mean computed from that sample is the only measure of precision for the estimate of the mean.

1. Describe the *data distribution* for individuals sampled from the Non-Normal population. Your description should include the center, spread (use the standard deviation and the range to describe spread), and shape of the distribution.

    Compare the center, spread, and shape of the Non-Normal population data distribution to the empirical sampling distributions of the mean for samples of n = 5, 15, 40, and 100 individuals from this population. How are each of the sampling distributions for the mean different from, or similar to, the Non-Normal population data distribution?

2. Describe the *data distribution* for individuals sampled from the Normal population, including the distribution center, spread, and shape.

    Compare the center, spread, and shape of the Normal population data distribution to the empirical sampling distributions of the mean for samples of n = 5, 15, and 40 individuals from this population. How are each of the sampling distributions for the mean different from, or similar to, the Normal population data distribution?

3. Describe how the comparisons of data distributions and sampling distributions of the mean you made for #1 and #2 demonstrate the Central Limit Theorem (CLT).

Name _____   Date _____

Describe how the descriptions of the sampling distributions of the mean you made for #1 and #2 demonstrate the Law of Large Numbers (LLN).

4. It is a universal rule in the practice of science that larger sample sizes are always better than smaller sample sizes. Based on your answers to questions 1 to 3 above, explain the advantages of large sample sizes.

5. Compare the standard deviations of the data distributions for the Normal and Non-Normal populations. Which of these two populations had greater variability among individuals?

Compare the standard deviation of the means for repeated samples of $n = 15$ taken from the Normal population to the standard deviation of the means for repeated samples of $n = 15$ taken from the Non-Normal population. How does the level of variation in the population affect the precision of sample means? Make similar comparisons for other sample sizes.

If you wanted to attain the same precision for estimates of the mean for the Non-Normal population as you attained with samples of $n = 15$ from the Normal population, how would you need to modify the sampling from the Non-Normal population?

6. Exploring the random variation of sample means based on simulations of repeated sampling from the same population is similar to exploring the probability of getting heads/tails outcomes by flipping a coin many times. Each independent repetition of taking a sample of n individuals from the same population is analogous to flipping a coin once. The Law of Large Numbers tells us that the more repetitions of the random process we observe, the better will be our understanding of the probability distribution. In this simulation you have generated 1,000 repeated samples from each of the populations, and computed a mean for each. The mean and standard deviation of 1,000 sample means computed for any specific sample size represent empirical estimates of the true mean (μ) and the true standard deviation of the mean ($\sigma_{\bar{x}}$) for the sampling distribution of the mean for that sample size. These values are listed in the Descriptive Statistics tables you produced for each population and sample size. Enter these standard deviations of means (S.D. $\bar{x}$'s) from 1,000 repeated samples in the appropriate spaces of the table in question 6.a. below.

The true population mean and standard deviation for the Normal population are $\mu = 100$, $\sigma = 20$. The mean μ and standard deviation σ for the Non-Normal population depend on the statistics software you used for the simulation. Enter the values specified in the tutorial here: $\mu =$ _____. $\sigma =$ _____.

a. Given these population standard deviations (σ's), use formula [1] in the Introduction for the Chapter 6 exercise to compute values of the "Theoretical" standard deviation of the mean ($\sigma_{\bar{x}}$) for sampling distributions of the mean based on sample sizes $n = 5$, 15, and 40 for both the Non-Normal and Normal populations.

	Non-Normal Population			Normal Population		
	($n=5$)	($n=15$)	($n=40$)	($n=5$)	($n=15$)	($n=40$)
S.D. of $\bar{x}$'s	___	___	___	___	___	___
$\sigma_{\bar{x}} =$	___	___	___	___	___	___

    b. Compare the theoretical values for $\sigma_{\bar{x}}$ to the Empirical estimates (S.D. of $\bar{x}$'s) derived from repeated samples. Are the values similar? Why would we *not* expect the empirical values to be exactly equal to the theoretical values?

    c. How might this comparison of empirical vs. theoretical values for the standard deviation of the mean be different if it had been based on the means and standard deviations for 50 means from repeated samples, rather than the 1,000 means you generated in your simulation? Explain how this outcome is related to the Law of Large Numbers.

10. Suppose that the governmental health organization sponsoring the clinical trial described in Problem 3 decided to recommend widespread use of the AZT treatment only if mother-to-infant HIV transmission rate could be reduced by at least 30%. That is, a decrease from $P = 0.3$ to 0.21 is the smallest difference of practical importance. Determine the power of this clinical trial with $n = 300$ subjects to detect a 30% decrease in transmission rate using a one-tailed test with acceptable Type I error rate $\alpha = 0.05$ (i.e., the health organization will conclude that AZT reduced transmission rate if the p-value from their one-tailed test of significance is ≤ 0.05).

a. Scale the X-axis for the sampling distribution of $\hat{p}$ under the premise that AZT has no effect on HIV transmission rate and $P = 0.3$. Shade the area under the curve that corresponds to the probability of obtaining $\hat{p}$ values that would provide sufficient evidence to conclude that the AZT treatment *reduces* transmission rate (based on a one-tailed test with $\alpha = 0.05$).

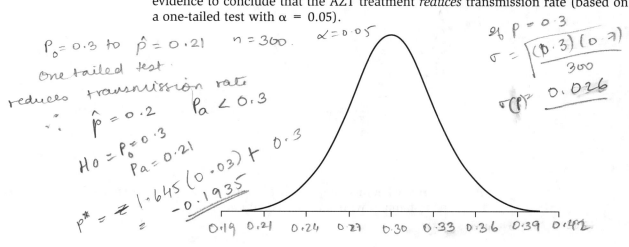

$P_0 = 0.3$ to $\hat{p} = 0.21$ $n = 300$. $\alpha = 0.05$

one tailed test.

reduces transmission rate

$\therefore \hat{p} = 0.2$ $P_a < 0.3$

$H_0 = P_0 = 0.3$

$P_a = 0.21$

$p^* = z \cdot 1.645 (0.03) + 0.3$

$= -0.1935$

$=$

if $p = 0.3$

$\sigma = \sqrt{\dfrac{(0.3)(0.7)}{300}}$

$\sigma(\hat{p})^2 \quad 0.026$

0.19 0.21   0.24   0.27   0.30   0.33 0.36   0.39   0.42

Note: The boundary of this "rejection region" on the X-axis would be -1.645 standard deviation units (negative tail $Z_{0.05}$) below P_0.

b. Compute the critical value $\hat{p}^*$ that corresponds to the boundary line between "fail to reject H_0" and "reject H_0," based on a one-tailed test of significance with $\alpha = 0.05$.

Name _____   Date _____

c. Scale the X-axis of the sampling distribution of $\hat{p}$ assuming that the AZT treatment has the minimum important effect on transmission rate of HIV (P_a = 0.21). You must specify $E(\hat{p})$ and recompute $\sigma_{\hat{p}}$ using this value for P. Shade the area under the curve that corresponds to the probability of obtaining a $\hat{p}$ value that would provide sufficient evidence to conclude that the AZT treatment reduced transmission rate, based on a one-tailed test with α = 0.05. *Note:* You determined the boundary of this area ($\hat{p}^*$) in part (b).

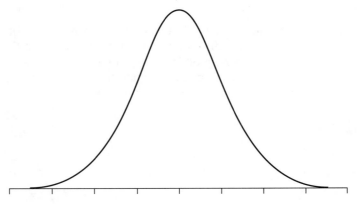

d. Describe what this sampling distribution in part (c) represents. Use terms that someone unfamiliar with statistics could understand.

e. Use the sampling distribution described in part (c) to determine the probability $P[\hat{p} \leq \hat{p}^*$ if P = 0.21]. Compute the Z-value associated with $\hat{p}^*$ and look up this value in the Standard Normal Table. This probability is the power of this clinical trial.

f. Interpret the probability you determined for part (e). Use words that would be understandable to someone who has not studied statistics.

g. Given the value for power computed above, compute the Type II error rate. What does this number mean? Use terms understandable to someone who has not had a statistics course. *Enter the results of Problem 10 in the table located in Problem 14.*

11. Suppose the researchers in Problem 10 were able to include only $n = 100$ pregnant women in their study, but all other aspects of the study were the same as before. Recompute the power of this study to detect a 30% reduction in HIV transmission rate using a one-tailed test of significance and an allowable Type I error rate of $\alpha = 0.05$.

a. Scale the X-axis for the sampling distribution of $\hat{p}$ under the premise that AZT has no effect on HIV transmission rate and $P_0 = 0.3$. Shade the area under the curve that corresponds to the probability of obtaining $\hat{p}$ values that would provide sufficient evidence for the researchers to conclude that the AZT treatment was effective.

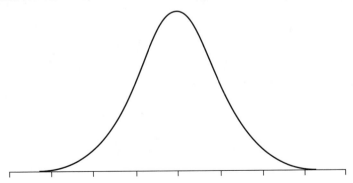

b. Compute the critical value for $\hat{p}^*$ that corresponds to this boundary line between the "fail to reject H_0" and "reject H_0," based on a one-tailed test with $\alpha = 0.05$.

c. Scale the X-axis of the sampling distribution of $\hat{p}$ assuming that the AZT treatment has the minimum important effect on HIV transmission rate ($P_a = 0.21$). Shade the area under the curve that corresponds to the probability of obtaining a $\hat{p}$ value that would cause the researchers to conclude that the AZT treatment reduced transmission rate, $P[\hat{p} \leq \hat{p}^*]$.

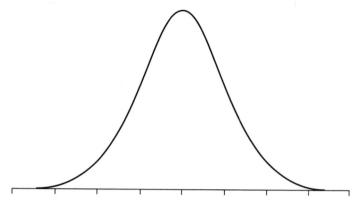

d. Determine the probability $P[\hat{p} \leq \hat{p}^*$ if $P = 0.21]$ by computing the Z-value associated with $\hat{p}^*$ and looking up this value in the Standard Normal Table. This probability is the power of this clinical trial with $n = 100$. *Enter the results for Problem 11 in the table located in Problem 14.*

Name _____ Date _____

12. For each of the following scenarios: (1) State the appropriate critical Z-value you would use to compute the critical sample proportion value $\hat{p}^*$ in step 1 of the power calculation. (2) State the appropriate P_a value for the minimum important difference you would use to compute the probability $P[\hat{p} \leq$ or $\geq \hat{p}^*$ if $P = P_a]$ in step 2 of the power calculations.

a. The investigator wants to determine if a new treatment for cancer *reduces* the proportion of patients who die within five years of diagnosis from the current level of $P = 0.7$. She would consider a 20% relative decrease in five-yr mortality rate to be the minimum effect size required for her to implement the new treatment. She is willing to accept a Type I error rate of $\alpha = 0.05$, using a one-tailed test of significance because she is only interested in whether or not the treatment reduces five-yr mortality rate.   *[handwritten: $\alpha = 0.05$ one tail test $\therefore$ $z_{0.05}$: negative]*

[handwritten left margin: $H_0: P = 0.7$; $H_a: P < 0.7$; one tailed test; $P_a < P_0$; $\therefore$ -ve]

[handwritten above 'a new treatment': 0.2]

Critical Z-value = __−1.645__ P_a = __$0.7 \times 0.2 = 0.14$ $\therefore$ $0.7 - 0.14$__
[handwritten: $P_a = 0.56$; $= 0.56$]

b. Consider the same scenario as in part (a), but now the investigator considers a 10% relative decrease in five-yr mortality rate to be the minimum important effect size.

Critical Z-value = _____ P_a = _____

c. Consider the same scenario as in part (a), but now the investigator decides she should use a two-tailed test of significance.

Critical Z-value = _____ P_a = _____

d. Consider the same scenario as in part (a), but now the investigator is willing to accept a Type I error rate of only $\alpha = 0.01$.

Critical Z-value = _____ P_a = _____

e. Consider the same scenario as in part (a), but now the researcher frames the question in terms of *survival rate* rather than mortality rate. That is, she wants to determine if the new treatment results in a 20% *increase* (above the current level of $P = 0.3$) in the proportion of patients who survive for at least five years after diagnosis of their cancer.

Critical Z-value = _____ P_a = _____

13. Suppose the researchers described in Problem 10 decided that it would be more appropriate if they analyzed their results using a two-tailed test of significance (with overall $\alpha = 0.05$). Recompute the power of the study (with $n = 300$) to detect a 30% reduction in mother-to-infant HIV transmission rate.

a. Scale the X-axis for the sampling distribution of $\hat{p}$ under the premise that AZT has no effect on HIV transmission rate ($P = 0.3$). Shade the area under the curve that corresponds to $\hat{p}$ values that would provide sufficient evidence to conclude that the AZT treatment reduces HIV transmission rate, based on a two-tailed test with $\alpha = 0.05$.

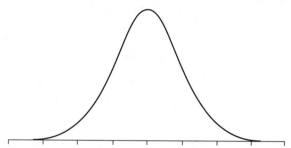

b. Compute the critical value for $\hat{p}^*$ that corresponds to this boundary line between the "fail to reject H_0" and "reject H_0," based on a two-tailed test with $\alpha = 0.05$. If the true value of P differs from P_0, it can be less than *or* greater than 0.3, but not both. Since the investigator is interested in *reducing* HIV transmission rate, do this calculation for the lower boundary between "reject" and "fail to reject" H_0.

c. Scale the X-axis of the sampling distribution of $\hat{p}$ assuming that the AZT treatment has the minimum important effect on HIV transmission rate ($P_a = 0.21$). Shade the area under the curve that corresponds to the probability of obtaining $\hat{p}$ values that would cause the researchers to conclude that the AZT treatment reduced HIV transmission rate, using a two-tailed test with $\alpha = 0.05$, $P[\hat{p} \leq \hat{p}^*]$.

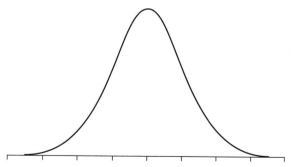

d. Determine the probability $P[\hat{p} \leq \hat{p}^*$ if $P = 0.21]$. This probability is the power of this clinical trial, based on a two-tailed test of significance with $\alpha = 0.05$. *Enter the results for Problem 13 in the table located in Problem 14.*

14. Answer the following questions based on the compilation of results from power calculations for variations on the clinical trial to test the effectiveness of AZT for reducing mother-to-infant HIV transmission in the table below.

	Sample Size (n)	(α) Type I Error Rate	Type of Test	$\sigma_{\hat{p}}$ for $P = P_0$	MSD = ($\hat{p}^* - P_0$)	MID = ($P_a - P_0$)	$(1 - \beta)$ Power	(β) Type II Error Rate
Pr10	300	0.05	1-tail	_____	_____	0.09	_____	_____
Pr11	**100**	0.05	1-tail	_____	_____	0.09	_____	_____
Pr13	300	0.05	**2-tail**	_____	_____	0.09	_____	_____

MSD = minimum significant difference; MID = minimum important difference.

a. Compare the results from Problems 10 and 11. What is the effect of decreasing sample size on the power of a study? Explain based on how decreasing sample size affected the sampling distributions and the magnitude of the *minimum significant* effect size ($\hat{p} - P_0$).

b. Compare the results for Problems 10 and 13. What do you conclude with regard to the difference in power between one-tailed vs. two-tailed tests? Explain this difference.

Confidence   **15.** In Problem 1, 560 of 1,000 ($\hat{p}$ = 0.56) women who experienced severe morning sick-
Intervals   ness gave birth to girls. The investigators concluded there was strong evidence that
women who have severe morning sickness are more likely to give birth to girls than
the overall population. They want to quantify the "effect size" by computing a confidence interval.

a. Compute the 95% confidence interval for P = the true proportion of female babies
born to women with severe morning sickness, based on these study results.

b. Interpret the meaning of this 95% confidence interval. Use terms that could be understood by someone who has never studied statistics.

c. In Problem 2, a similar study found that 179 of 320 ($\hat{p}$ = 0.56) women who experienced severe morning sickness gave birth to girls. Compute the 95% confidence
interval for P = the true proportion of female babies *born to women with severe
morning sickness*, based on these study results.

d. How did changing the sample size affect the margin of error, width, and error rate
of the confidence interval? Explain these changes using plain language.

Margin of error: _____   Width: _____   Error rate: _____

Explanation: _____

e. Compute the 99% confidence interval for P based on the study described in part (c).

f. Compare the 99% confidence interval computed for part (e) with the 95% confidence interval computed for part (c). How does increasing the confidence level affect the margin of error, width, and error rate of the confidence interval? Explain why
this is so. Use plain language.

Margin of error: _____   Width: _____   Error rate: _____

Explanation: _____

16. Suppose the researchers studying the proportion of girl children born to women who experience severe morning sickness wanted a 95% confidence interval for *P* with a margin of error of only ± 0.02 (m^*). What sample size (n^*) would be required?

   a. Do this calculation assuming the researchers had no knowledge about the value of *P*.

   b. Redo the calculation using the information $\hat{p}$ = 0.56 as an estimate of *P* that was obtained from a preliminary study.

   c. If these researchers wanted a 99% confidence interval with a margin of error of no larger than ± 0.02, how large a sample size would be needed? Assume no prior information.

   d. How does the desire for increased "confidence" affect the sample size required to attain the desired margin of error?

17. Having come to the conclusion that the AZT treatment reduced HIV transmission rate between mothers and their babies, the researchers described in Problem 3 now want to document the effect size. Given their results that 75 of 300 babies were HIV positive:

   a. Compute the 95% confidence interval for *P* = the true proportion of babies infected with HIV during birth when both mother and baby receive the AZT treatment.

   b. What are the margin of error and error rate for this confidence interval?

       Margin of error: _____   Error rate: _____

c. If the study had been based on a sample size of only 100 (instead of 300) and 25 babies had tested HIV positive, how would you expect this to change the margin of error and the error rate of this 95% confidence interval, if at all? Explain the conceptual reason for your answers.

Margin of error: _____   Error rate: _____

d. The confidence interval computed for part (a) does not include the value 0.3, which is the HIV transmission rate from mothers to infants when they do *not* receive the HIV treatment. What can you conclude from this observation?

e. What are the assumptions that must be fulfilled for the confidence intervals you computed for Problems 15 to 17 to be valid? For each assumption, *state why* that assumption is necessary, and *state how* you would determine whether or not the assumption is fulfilled.

f. Compute the 99% confidence interval for *P* based on the study described in part (a).

g. Compare the 99% confidence interval computed for part (f) with the 95% confidence interval computed for part (a). How does increasing the confidence level affect the margin of error, width, and error rate of the confidence interval? Explain why this is so. Use plain language.

Margin of error: _____   Width: _____   Error rate: _____

18. Suppose the researchers studying the effectiveness of the new AZT treatment wanted a 95% confidence interval for P with a margin of error of ± 0.03 (m^*). What sample size (n^*) would be required?

a. Do this calculation assuming the researchers had no knowledge about the value of P.

b. Redo the calculation using the information $\hat{p} = 0.25$ as an estimate of P that was obtained from a preliminary study.

c. Given the "life-and-death" aspects of this research and the high cost of AZT, suppose these researchers wanted a 95% confidence interval with a margin of error of no larger than ± 0.01. How large a sample size would be needed? Assume no prior information.

d. How does the desire for a narrower, more precise confidence interval affect the sample size required to attain the desired margin of error?

Supplemental Problems for Chapter 7

19. Suppose that a researcher in Thailand wants to repeat the study of AZT effect on HIV transmission from mother to infant at the time of birth described in Problem 3. She can only afford to do the study with a sample of 50 women, due to the high cost of the drug. Suppose that of the women and children who were given AZT, HIV was transmitted to the infant in 10 cases ($\hat{p}$ = 0.2). Does this study provide sufficient evidence to conclude that the AZT treatment reduced the mother-to-infant HIV transmission rate from P = 0.3?

a. State the appropriate Null and Alternative hypotheses.

H_0: _____ H_a: _____

b. Describe the sampling distribution of $\hat{p}$ for this study if the Null hypothesis is true. Scale the X-axis of the distribution in accordance with this description. Shade the area under the curve that corresponds to the p-value for this test of significance.

Center: $E(\hat{p})$ = _____

Spread: $\sigma_{\hat{p}}$ = _____

Shape: _____

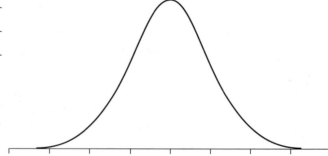

c. Compute the Z_{test} statistic, determine the p-value, interpret this probability, and state your conclusion with regard to the question.

d. What is the effect size of this study?

(1) Absolute effect size: _____

(2) Relative effect size: _____

e. This study had a larger effect size than that of the Problem 3 study $[(\hat{p} - \boldsymbol{P}_0) = 0.05]$, yet provided weaker evidence that AZT reduces HIV transmission from mothers to infants. Explain how this can be.

20. Given the "life-and-death" nature of the clinical trial to test the effectiveness of using AZT to reduce transmission of HIV from mothers to infants, suppose the researchers wanted to be especially careful about Type I error. They decide they will not conclude the AZT treatment is effective unless the p-value is ≤ 0.01. Recompute the power of this study, with $n = 300$ subjects, to detect a 30% decrease in HIV transmission with $\alpha = 0.01$.

a. Scale the X-axis for the sampling distribution of $\hat{p}$ under the premise that AZT has no effect on HIV transmission rate and $\boldsymbol{P}_0 = 0.3$. Shade the area under the curve that corresponds to the probability of obtaining $\hat{p}$ values that would provide sufficient evidence to conclude that the AZT treatment reduced HIV transmission rate, *based on a one-tailed test with $\alpha = 0.01$.*

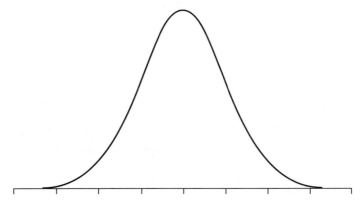

b. Compute the critical value for $\hat{p}^*$ that corresponds to this boundary line between the "fail to reject H_0" and "reject H_0 regions along the X-axis."

c. Scale the *X*-axis of the sampling distribution of $\hat{p}$ assuming that the AZT treatment has the minimum important effect on HIV transmission rate ($\boldsymbol{P}_a = 0.21$). Shade the area under the curve that corresponds to the probability of obtaining $\hat{p}$ values that would provide sufficient evidence to conclude that the AZT treatment reduced transmission rate, with one-tailed $\alpha = 0.01$.

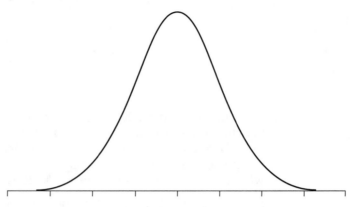

d. Determine the probability $P[\hat{p} \leq \hat{p}^*$ if $\boldsymbol{P} = 0.21]$. This is the power of this clinical trial with $\alpha = 0.01$.

e. Compare the results for Problem 10 and Supplemental Problem 20. What is the effect of decreasing the acceptable Type I error rate (α) on power? Did the investigators decrease the *overall* probability that their conclusion would be wrong ($\alpha + \beta$)? Explain.

21. AZT treatment is expensive, and this is problematic for poor countries with limited health care budgets. Money spent on AZT to prevent mother-to-infant HIV transmission might be used to improve primary health care for thousands of people. In this context, public health officials decided that the AZT treatment should reduce HIV transmission at least 30% before the cost/benefit analysis would favor providing the AZT treatment. In richer countries that can better afford the cost of AZT, public health officials could have decided that a 20% reduction in HIV transmission from mothers to infants would be sufficient to warrant implementing the treatment. What is the power of the HIV clinical trial with n = 300 to detect a 20% reduction in HIV transmission, using a one-tailed test with α = 0.05?

a. Scale the X-axis for the sampling distribution of $\hat{p}$ under the premise that AZT has no effect on HIV transmission rate and P_0 = 0.3. Shade the area under the curve that corresponds to the probability of obtaining $\hat{p}$ values that would provide sufficient evidence to conclude that the AZT treatment was effective, based on a one-tailed test with α = 0.05.

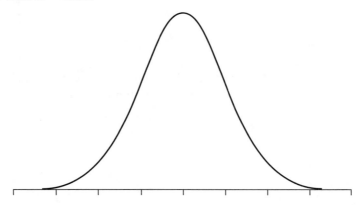

b. Compute the critical value for $\hat{p}^*$ that corresponds to this boundary line between "fail to reject H_0" and "reject H_0," based on a one-tailed test with α = 0.05.

c. Scale the *X*-axis of the sampling distribution of $\hat{p}$ assuming that the AZT treatment has the minimum important effect of reducing HIV transmission rate by 20%. Shade the area under the curve that corresponds to the probability of obtaining $\hat{p}$ values that would cause the researchers to conclude that the AZT treatment was effective.

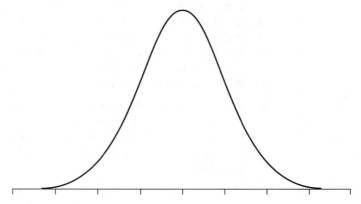

d. Determine the power of this clinical trial to detect a 20% reduction in HIV transmission rates from mother to infant.

e. Compare the results for Problem 10 and Supplemental Problem 21. What is the effect of decreasing the size of the minimum important difference on the power of a test of significance? Explain why you think this is so.

22. For each of the scenarios below, state the appropriate conclusion with regard to the original question in a format that would be appropriate for a Results section of a scientific report or paper. *Explain the logic behind each of your conclusions.*

a. A public health research wants to know if the proportion of African-American women in New York City who have health insurance is less than that for the overall U.S. population ($P = 0.65$). She does a randomized, unbiased survey of $n = 1,000$ African-American women and determines that $\hat{p} = 0.45$ of the women have health insurance. The p-value from a test of significance for this difference is $p < 0.0002$.

b. A plant ecologist wonders if bees are differentially attracted to plants with blue flowers or to plants with yellow flowers. If so, plants with the more attractive flower color would be better pollinated, produce more seeds, and become more abundant relative to the other flower color. Based on the genetics of flower color determination, the proportion of individuals with blue flowers is expected to be $P = 0.25$, but only if there is no natural selection for or against this flower color. The ecologist obtains a randomized, unbiased sample of 300 plants, of which 63 ($\hat{p} = 0.21$) have blue flowers. The p-value from a test of significance for this difference is $p = 0.0548$.

c. Suppose the p-value from the plant ecologist's study of the proportions of flowers with blue vs. yellow flowers was $p = 0.12$ and the other aspects of the study were the same as before.

23. Suppose the genetics student described in Problem 6 wanted to compute a confidence interval for the proportion of offspring produced by the controlled mating that have long wings and gray bodies. She observed 149 out of 241 ($\hat{p}$ = 0.62) offspring exhibited this combination of traits.

a. Compute the 95% confidence interval for P = the true proportion of offspring produced by this controlled mating that have long wings and gray bodies.

b. What are the margin of error and error rate for this confidence interval?

Margin of error: _____   Error rate: _____

c. If the study had been based on a sample size of only 100 and 62 offspring had exhibited this combination of traits, how would you expect this to change the margin of error and the error rate of this 95% confidence interval, if at all? Explain your answers.

Margin of error: _____   Error rate: _____

Explanation: _____

d. Based on the results used to compute the confidence interval described in part (a) above, compute the 90% confidence interval for P = the true proportion of offspring from this controlled cross that have long wings and gray bodies.

e. How does reducing the confidence level influence the following characteristics of the confidence interval?

Margin of error: _____   Error rate: _____

f. What aspect of the confidence interval reflects the *precision* of the estimate of P, the confidence level, the error rate, or the margin of error?

Study Problems for Chapter 7

1. A student uses controlled mating to produce a colony of fruit flies that should all have a heterozygous genotype. However, while transferring flies to new culture bottles, she thinks that some flies from the homozygous dominant culture might have gotten mixed into her heterozygous genotype culture. To test the purity of her heterozygous culture, she mates a sample of flies from this culture with flies that are homozygous recessive. If the flies in the colony are all heterozygous, then $P = 0.5$ of the offspring from this mating should exhibit the recessive trait. If some flies in the colony are homozygous for the dominant trait, the proportion of flies that exhibit the recessive trait will be less than 0.5. Suppose the outcome was 81 out of 180 flies in the sample that exhibited the recessive trait.

   a. State the appropriate Null and Alternative hypotheses appropriate for this study.

      H_0: _____ H_a: _____

   b. Scale the X-axis of the sampling distribution of $\hat{p}$ under the assumption that the colony is pure heterozygous individuals ($P = 0.5$), given the number of offspring produced is $n = 120$. Shade the area under the curve that corresponds to the p-value of a one-tailed test of significance.

      $E(\hat{p})$ = _____

      $\sigma_{\hat{p}}$ = _____

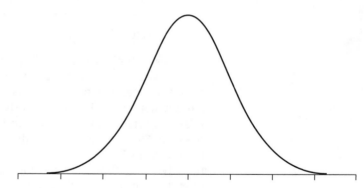

   c. Using this sampling distribution, perform a test of significance for the student's result. Compute the Z_{test} statistic, determine the p-value, and state your conclusion with regard to the student's original question.

d. Compute the 95% confidence interval for **P** based on this sample data.

e. Suppose the student repeats this test and gets $n = 240$ offspring, of which 108 exhibit the homozygous recessive trait. Compute the 95% confidence interval for **P**.

f. Compare the 95% confidence intervals computed for parts (d) and (e). How does increasing the sample size affect the width of this confidence interval? How does increasing sample size affect the error rate of this confidence interval?

Width: _____

Error rate: _____

2. Sclerosis of the aortic heart valve has long been observed by doctors in their patients over 65 years old, but this condition was always thought to be harmless. Although the condition itself has never been associated with health problems, a researcher wants to study whether it might be an early warning marker of increased risk of heart disease-related mortality. Suppose that in the United States population the proportion of people over 65 who die of heart disease is $P = 0.35$. The researcher compiles medical records from 400 people who had a test that would have identified sclerosis of the aortic valve sometime between the ages of 65 and 68. Of these 400 people, 160 subsequently died of heart disease ($\hat{p} = 0.4$). Perform a test of significance to address the scientific question "Do patients over 65 years old with sclerosis of the aortic valve differ from the general population of people who are over 65 with regard to risk of heart disease–related mortality?"

a. State the Null and Alternative hypotheses for this test.

H_0: _____ H_a: _____

b. Scale the X-axis of the sampling distribution of $\hat{p}$, assuming that people with sclerosis of the aortic valve die of heart disease at the same rate as the general population who are over 65 years old. Shade the area under the curve that corresponds to the *p*-value for the appropriate test of significance for the observed sample proportion ($\hat{p} = 0.4$).

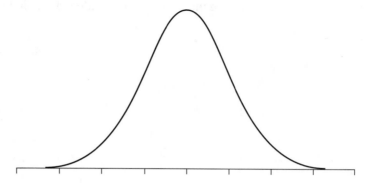

c. Compute the Z_{test} statistic and determine the *p*-value.

d. Explain what this *p*-value represents. Use terms understandable to someone who has not had a statistics course.

e. State your conclusion with regard to the researcher's original question.

f. At this point the *p*-value has another interpretation. Explain.

g. Compute the 95% confidence interval for P = the proportion of people over 65 with sclerosis of the aortic valve who die of heart disease. Interpret this 95% confidence interval in terms understandable by a nonstatistician.

h. Suppose that the researcher decides that if the proportion of people with sclerosis of the aortic valve who die of heart disease is $P \geq 0.45$, this would be a sufficiently large effect size to recommend that tests for this condition be used as an early warning sign for heart disease. (People with this condition might be instructed to take extra precautions against heart disease.) What is the power of the researcher's study, with n = 400 individuals, to detect this increased risk of heart disease–related mortality with a two-tailed test with α = 0.05?

Step 1: Scale the X-axis under the Null hypothesis. Determine the critical value $\hat{p}^*$ for a two-tailed test with α = 0.05.

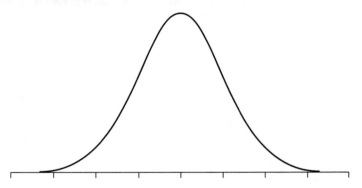

Step 2: Scale the X-axis under the Alternative hypothesis. Determine power = $P[\hat{p} \geq \hat{p}^* \text{ if } P = 0.45]$.

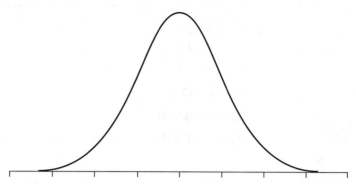

Answer Key for Chapter 7 Study Problems

1. **a.** $H_0: P = 0.5$
   $H_0: P < 0.5$

   b. $E(\hat{p}) = 0.5$

   $\sigma_{\hat{p}} = 0.0373$

   Shape: **Normal**

   c. $P[\hat{p} \le 0.45$ if $P = 0.5$ and $\sigma_{\hat{p}} = 0.0373] = P[Z \le (0.45 - 0.5) / 0.0373]$

   $\qquad\qquad\qquad\qquad\qquad\qquad\qquad\qquad\;\; = P[Z \le -1.34]$

   $\qquad\qquad\qquad\qquad\qquad\qquad\qquad\qquad\;\; = \mathbf{0.0901}$

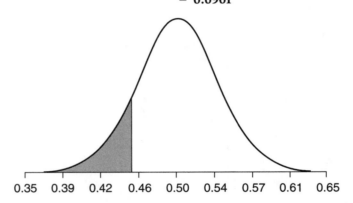

   Conclusion: The results suggest that the proportion of homozygous recessive flies produced by this culture is lower than it would be if the culture were pure. That is, these results indicate that the culture was contaminated by homozygous dominant flies, but more data are required to verify this conclusion.

   d. $\hat{p} \pm Z_{0.025} \sqrt{\hat{p}(1 - \hat{p})/n} = 0.45 \pm 1.96 \sqrt{0.45\,(0.55) / 180}$

   $\qquad\qquad\qquad\qquad\qquad\quad\; = 0.45 \pm 1.96\,(0.0371)$

   $\qquad\qquad\qquad\qquad\qquad\quad\; = \mathbf{0.45 \pm 0.0727}$

   e. $\hat{p} \pm Z_{0.025} \sqrt{\hat{p}(1 - \hat{p})/n} = 0.45 \pm 1.96 \sqrt{0.45\,(0.55) / 240}$

   $\qquad\qquad\qquad\qquad\qquad\quad\; = 0.45 \pm 1.96\,(0.0321)$

   $\qquad\qquad\qquad\qquad\qquad\quad\; = \mathbf{0.45 \pm 0.0629}$

   f. Increasing sample size decreases the margin of error and the width of the confidence interval because sample statistics computed from larger samples exhibit less random sampling variation.

   Increasing sample size has *no effect* on the error rate of the confidence interval. Error rate ($\alpha = 0.05$) was fixed when the investigator decided on a 95% confidence level.

2. a. $H_0: P = 0.35$

      $H_0: P \neq 0.35$

  b. $E(\hat{p}) = 0.35$

      $\sigma_{\hat{p}} = 0.0238$

      Shape: **Normal**

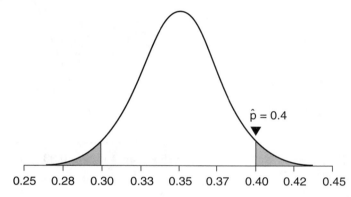

  c. $P[\hat{p} \geq 0.4 \text{ if } P = 0.35 \text{ and } \sigma_{\hat{p}} = 0.0238] = P[Z \geq (0.4 - 0.35) / 0.0238]$

                                             $= P[Z \geq +2.10]$

                                             $= \mathbf{0.0179 \times 2 = 0.0358}$

  d. The *p*-value is the probability of obtaining the observed difference (0.4 − 0.35 = 0.05) in either direction (+ or −) due only to random variation (if there really is no difference in risk of heart disease between people with or without sclerosis of the aortic valve).

  e. *Conclusion:* There is strong evidence that a greater proportion of people over 65 years old who have sclerosis of the aortic valve die of heart disease than those who do not have this condition (*p* = 0.0358).

  f. Suppose you have concluded that there is a difference in proportion who die of heart disease between people with/without sclerosis of the aortic valve. The *p*-value is now the probability that this conclusion is *incorrect*. In this case, there is a probability 0.0358 that the observed difference is due only to random variation, and so, this conclusion constitutes a Type I error.

  g. 95% confidence interval for $P = \hat{p} \pm Z_{0.025} S_{\hat{p}}$

                              $= 0.4 \pm 1.96 \sqrt{0.4\,(0.6)/400} = 0.4 \pm 1.96\,(0.0245)$

                              $= 0.4 \pm 0.0480$

Because 95% of all confidence intervals computed by this method include the value of the true population proportion **P**, we are 95% confident that the true proportion of people who are over 65 years old and have sclerosis of the aortic valve that subsequently die of heart disease is within the range 0.4 ± 0.0480.

h. Step 1

$$\hat{p}^* = +Z_{0.025} (\sigma_{\hat{p}}) + P_0$$

$$= +1.96 \, (0.0238) + 0.35$$

$$= 0.397$$

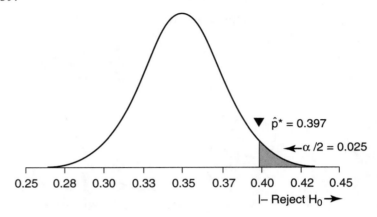

Step 2

$$P[\hat{p} \geq 0.397 \text{ if } \mathbf{P} = 0.45] \quad = P[Z \geq (0.397 - 0.45) \, / \, 0.0249]$$

$$= P[Z \geq -2.13]$$

$$= 1 - P[Z \leq -2.13]$$

$$= 1 - 0.017$$

$$= 0.9834$$

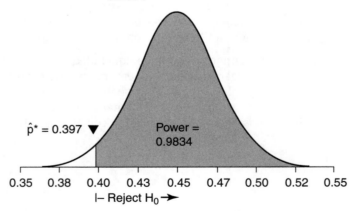

Chapter 8 Supplemental Problem

5. Concerns have been raised that women who receive an epidural (a spinal anesthetic) to relieve the pain associated with childbirth are more likely to require a Caesarian section (C-section) than women who receive a narcotic pain reliever. Some believe that an epidural may slow labor and inhibit the mother's ability to push the baby through the birth canal. A study compared the proportion of women who required a C-section between a group of women who received an epidural (e) and a group who received a narcotic pain reliever (n). Of 1,183 women who received an epidural, 97 delivered their baby by C-section. Of 1,186 women who were given a narcotic for their labor pain, 67 delivered by C-section. Perform a two-sample Z-test to determine if these results provide sufficient evidence to conclude that the proportion of women who require a C-section differs according to whether the women had received an epidural or a narcotic pain reliever.

   a. State the appropriate H_0 and H_a.

   H_0: _____ H_a: _____

   b. Compute the pooled estimate of the proportion of women who required delivery by C-section if the Null hypothesis is true, and compute $S_{(\hat{p}e-\hat{p}n)}$, the standard error of the difference $(\hat{p}_e - \hat{p}_n)$.

   c. Describe the sampling distribution of $(\hat{p}_e - \hat{p}_n)$ and scale the X-axis under the premise that the Null hypothesis is correct. Shade the area under the curve that corresponds to the p-value from the test of significance for this question.

   $E(\hat{p}_e - \hat{p}_n)$ = _____

   $S_{(\hat{p}e-\hat{p}n)}$ = _____

   Shape: _____

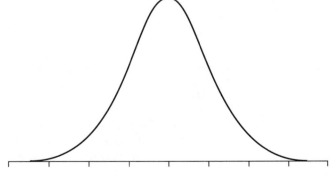

d. Compute the Z_{test} statistic and determine the *p*-value of the test statistic.

e. Explain the interpretation of this *p*-value in the context of this study. Use terms understandable by someone who has not studied statistics.

f. State your conclusion with regard to the original question. Use a format appropriate for a scientific paper.

g. Compute the 95% confidence interval for the true difference in the proportion of women who require a C-section between those who receive an epidural and those who receive narcotic pain relievers $(P_e - P_n)$.

h. Explain why we can't just say that this difference in the proportion of women who require a C-section is $(\hat{p}_e - \hat{p}_n) = (0.0820 - 0.0565) = 0.0255$. Use terms understandable to someone who has not studied statistics.

    i. Explain the interpretation of this 95% confidence interval in this context. Use terms understandable to someone who has not studied statistics.

Chapter 8 Study Problems

1. It has been widely documented that the proportion of women who deliver their babies by Caesarian section (rather than natural, vaginal birth) increased during the latter part of the twentieth century. Some women prefer that their children be born naturally, and they are concerned that doctors may be performing C-sections at the first hint of difficulty in the birthing process so as to avoid later malpractice lawsuits should something go wrong. Suppose that a woman wants to deliver her baby by natural birth. She has done some research and found that the proportion of births by C-section for the whole United States is 6% ($P = 0.06$). She asks her obstetrician how many C-sections he has done recently. He replies that 20 out of 230 deliveries he performed in the past year were by C-section. Should the woman be concerned that her obstetrician is more likely than the national average to deliver by C-section?

    a. State the appropriate Null and Alternative hypotheses for this analysis.

        H_0: _____ H_a: _____

    b. Describe the sampling distribution for $\hat{p}$ appropriate for this analysis, and scale the X-axis under the premise that the Null hypothesis is correct. Shade the area under the curve that corresponds to the p-value from the appropriate test of significance for this question.

        $E(\hat{p})$ = _____

        $S_{\hat{p}}$ = _____

        Shape: _____

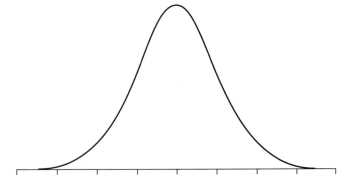

    c. Use this sampling distribution to determine the *p*-value for the sample proportion.

    d. State your conclusion with regard to the original question.

    e. Compute the 95% confidence interval for the true proportion of C-section births performed by this obstetrician.

2. Suppose that the pregnant woman wants to deliver her baby by natural birth but is afraid that hospital policies designed to minimize risk of malpractice lawsuits will require a C-section at the first sign of difficulty. She investigates if the proportion of births by C-section (vs. natural birth) differs between the two hospitals in her city. (If the hospitals differ, she will choose the hospital with the lowest proportion of C-section births.) Hospital A replies that during the past year 300 out of 3,100 deliveries were by C-section. Hospital B replies that 120 out of 1,130 deliveries in the past year were by C-section. Is this sufficient evidence to conclude that the hospitals differ with regard to the proportion of deliveries done by C-section?

a. State the appropriate H_0 and H_a.

H_0: _____ H_a: _____

b. Compute the pooled estimate of the proportion of births by C-section if the Null hypothesis is true, and compute the standard error of the difference $(\hat{p}_A - \hat{p}_B)$, $S_{(\hat{p}A-\hat{p}B)}$.

c. Describe the sampling distribution of $(\hat{p}_A - \hat{p}_B)$ and scale the X-axis under the premise that the Null hypothesis is true. Shade the area under the curve that corresponds to the p-value from the appropriate test of significance for this question.

$E(\hat{p}_A - \hat{p}_B)$ = _____

$S_{(\hat{p}A-\hat{p}B)}$ = _____

Shape: _____

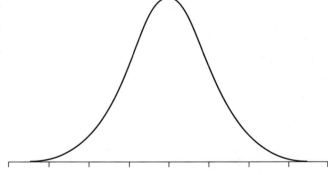

d. Compute the Z_{test} statistic and determine the *p*-value of the test statistic.

e. State your conclusion with regard to the original question, including which hospital you think the woman should go to if she wants to increase her chances of natural birth.

f. Compute the 95% confidence interval for the true difference in the proportion of births by C-section between these two hospitals ($P_A - P_B$).

Name _____   Date _____

Answer Key for Chapter 8 Study Problems

1. **a.** $H_0: P = 0.06$ $H_a: P \neq 0.06$

   b. $E(\hat{p}) = 0.06$

   $\sigma_{\hat{p}} = 0.0157$

   $\quad = \sqrt{(0.06)(0.94)/230}$

   Shape: **Normal**

   $0.06\,(230) = 13.8 > 10$

   $0.94\,(230) = 216 > 10$

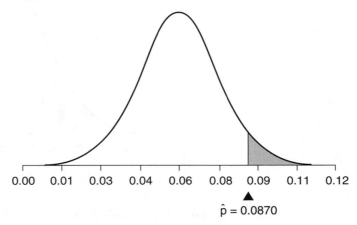

$\hat{p} = 0.0870$

   c. $\hat{p} = X/n = 20/230 = \mathbf{0.0870}$

   $Z_{\text{test}} = \dfrac{0.087 - 0.06}{0.0157} = \mathbf{1.72}$

   $P[\hat{p} \geq 0.087 \text{ if } \boldsymbol{P} = 0.06] = P[Z \geq +1.72] = 0.0427$

   Note: The woman wanted to determine only if her obstetrician was **more** likely than average to perform a c-section. Hence a one-tailed test was performed.

   d. *Conclusion:* The results indicate that this woman's obstetrician does a greater proportion of deliveries by C-section compared to the national rate of 0.06 ($p = 0.0427$).

   e. 95% confidence interval for $\boldsymbol{P}$ = the true proportion of C-section births by this obstetrician:

   $\hat{p} \pm Z_{0.025}\ \sqrt{\hat{p}(1 - \hat{p})/n} = 0.087 \pm 1.96\ \sqrt{0.087(0.913)/230}$

   $\qquad\qquad\qquad\qquad = 0.087 \pm 1.96\,(0.0186)$

   $\qquad\qquad\qquad\qquad = \mathbf{0.087 \pm 0.0364}$

2. a. H_0: $(P_A - P_B) = 0$ H_a: $(P_A - P_B) \neq 0$

b. $\hat{p}_{pool} = (300 + 120) / (3100 + 1130) = \mathbf{0.0993}$

$$S_{(\hat{p}A-\hat{p}B)} = \sqrt{(0.0993)(1 - 0.0993)[(1/3100) + (1/1130)]}$$

$$= \sqrt{(0.0993)(0.9007)(0.00121)}$$

$$= \mathbf{0.0104}$$

c. $E(\hat{p}_A - \hat{p}_B) = \mathbf{0}$

$S_{(\hat{p}A-\hat{p}B)}$ $= \mathbf{0.0104}$

Shape: **Normal**

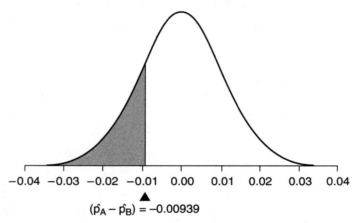

$(\hat{p}_A - \hat{p}_B) = -0.00939$

d. $\hat{p}_A = 300 / 3100 = \mathbf{0.0968}$ $\hat{p}_B = 120 / 1130 = \mathbf{0.1062}$

$(\hat{p}_A - \hat{p}_B) = -0.00939$

$P[(\hat{p}_A - \hat{p}_B) \leq -0.00939 \text{ if } (P_A - P_B) = 0] = P[Z \leq (-0.00939 / 0.0104)$

$= P[Z \leq -0.903]$

$= 0.1841 \times 2 = \mathbf{0.3682}$ (two-tailed p-value)

e. *Conclusion:* The proportion of births by C-section does not differ between these two hospitals ($p = 0.3682$).

f. $S_{(\hat{p}A-\hat{p}B)} = \sqrt{\dfrac{\hat{p}_A(1 - \hat{p}_A)}{n_A} + \dfrac{\hat{p}_B(1 - \hat{p}_B)}{n_B}}$

$= \sqrt{\dfrac{0.0968(0.9032)}{3100} + \dfrac{0.1062(0.8938)}{1130}}$

$= \mathbf{0.0106}$

95% confidence interval for $(P_A - P_B) = (\hat{p}_A - \hat{p}_B) \pm Z_{0.025}(S_{(\hat{p}A-\hat{p}B)})$

$= -0.00939 \pm 1.96 (0.0106)$

$= \mathbf{-0.00939 \pm 0.0208}$

Name _____   Date _____

11. The marketers of Ginkgo biloba nutritional supplements claim that the active ingredient in their product can enhance memory. To test this claim, a researcher gives a sample of $n = 30$ subjects a standard memory test, has them take the Ginkgo supplements for three months, and then retake the memory test to see if their scores increased. The data for this experiment are given below.

The Data

ID#	Pre-Test	Post-Test	ID#	Pre-Test	Post-Test	ID#	Pre-Test	Post-Test
1	73	76	11	59	68	21	64	64
2	55	58	12	54	53	22	44	46
3	73	73	13	60	66	23	59	59
4	58	60	14	67	72	24	67	67
5	54	60	15	62	56	25	52	54
6	56	63	16	61	67	26	63	74
7	60	55	17	67	66	27	64	71
8	74	77	18	56	59	28	56	56
9	76	84	19	77	83	29	76	81
10	62	75	20	69	71	30	69	68

- Enter the pre- and post-test data into a computer statistics program as two variables named **Pre** and **Post** (don't bother entering the ID #'s). Each variable should contain 30 data values. *Important:* You must maintain the pairing of the Pre and Post data values for each subject by having them on the same row in the two variables.

- The Normality assumption for the paired t-test pertains to the sampling distribution of $\overline{D}$, the mean of the individual difference values (D = Post – Pre). To determine if the Normality assumption is valid, you must first compute these difference values. Create a variable **Diff** that contains the differences (**Post – Pre**) for all of the study subjects.

- Produce summary statistics (sample size, mean, median, standard deviation) and graphics (boxplot, Normal quantile plot) for the D-values in variable **Diff**. Copy/Paste the output from these analyses into a single word processor page. *Below these analyses, type your assessment of whether or not the Normality assumption for the paired t-test is fulfilled.*

- Use your statistics software to perform a paired t-test on the Pre/Post data.

- Copy/Paste the t-test print-out to the word processor page, below the assessment of the Normality assumption. Append that page to your answers to questions about these analyses on the next page. Save the data file to diskette; you will further analyze these data in later problems.

a. State the appropriate Null and Alternative hypotheses for this test of significance.

H_0: _____ H_a: _____

b. Using the information in the question and results from summary statistics, scale the X-axis for the sampling distribution for $\overline{D}$ under the assumption that H_0 is true. Shade the area under the curve that corresponds to the p-value.

$E(\overline{D})$ = _____

$S_{\overline{D}}$ = _____

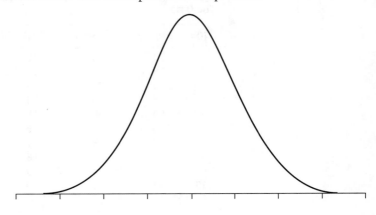

c. Explain what the standard error $S_{\overline{D}}$ represents.

d. Interpret the meaning of the p-value from the paired t-test in the context of this study.

e. Based on the results from the paired *t*-test, state your conclusion regarding the effectiveness of Ginkgo supplements for enhancing memory. *Important: Enter the results from this test of significance in the first line of the table in Problem 20.*

f. Having stated your conclusion, explain the second interpretation of the *p*-value.

g. Compute the 95% confidence interval for the mean difference in memory test scores associated with the Ginkgo supplement (μ_D). Use the exploratory data analysis results to obtain $\bar{D}$ and S_D.

h. Explain the interpretation of this confidence interval in this context. Use words understandable to someone who has not studied statistics.

12. Suppose that the average memory test score on the pre-test was 60 points out of a possible 100. The investigators think that if the Ginkgo supplement can increase scores on the memory test by 5% this would be sufficient evidence to claim that Ginkgo enhances memory. Determine the power of this study to detect an average increase in the memory test score of 3 points. [Minimum important effect size $= +0.05$ (60 pts) $= 3$] for a sample of $n = 30$ subjects, using a one-tailed test with $\alpha = 0.05$, given the data in Problem 11.

a. Scale the X-axis of the sampling distribution of $\overline{D}$ under the assumption of the Null hypothesis. Shade the area under the curve that represents the probability associated with $\overline{D}$ values that would provide sufficient evidence to reject the Null hypothesis.

$E(\overline{D}) = $ _____

$S_{\overline{D}} = $ _____

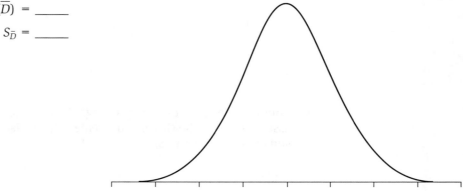

b. Compute the smallest value for $\overline{D}^{*}$ that would provide sufficient evidence to reject the Null hypothesis with a p-value ≤ 0.05.

c. Scale the X-axis of the sampling distribution of $\overline{D}$ if the seminar increases test scores an average of 3 points but does not change the standard deviation of the population σ. Shade the area under this curve that corresponds to the power of the test.

$E(\overline{D}) = $ _____

$S_{\overline{D}} = $ _____

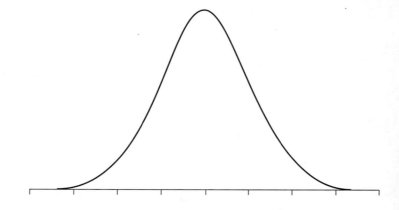

Chapter 9 Supplemental Problems

24. Given the following information on H_0 and H_a, sample size, and t-value from a test of significance, determine the probability associated with the specified t-values.

Hypotheses	n	t-Value	Probability
$H_0: \mu = 0$ $H_a: \mu > 0$	10	4.01	_____
$H_0: \mu = 0$ $H_a: \mu > 0$	15	-0.78	_____
$H_0: \mu = 0$ $H_a: \mu > 0$	45	1.68	_____
$H_0: \mu = 0$ $H_a: \mu > 0$	75	-0.48	_____
$H_0: \mu = 0$ $H_a: \mu < 0$	20	-1.34	_____
$H_0: \mu = 0$ $H_a: \mu < 0$	30	2.51	_____
$H_0: \mu = 0$ $H_a: \mu < 0$	60	-0.31	_____
$H_0: \mu = 0$ $H_a: \mu < 0$	90	0.98	_____
$H_0: \mu = 0$ $H_a: \mu \neq 0$	5	-1.07	_____
$H_0: \mu = 0$ $H_a: \mu \neq 0$	25	-2.30	_____
$H_0: \mu = 0$ $H_a: \mu \neq 0$	47	1.05	_____
$H_0: \mu = 0$ $H_a: \mu \neq 0$	56	4.37	_____

25. Given the following values for sample size, standard deviation of the data values, and acceptable error rate, compute the margin of error for confidence intervals for μ:

	Confidence Level (%)	Margin of Error
$n = 10$ $S = 15$ $\alpha = 0.05$	_____	_____
$n = 20$ $S = 25$ $\alpha = 0.01$	_____	_____
$n = 44$ $S = 12$ $\alpha = 0.10$	_____	_____
$n = 60$ $S = 40$ $\alpha = 0.05$	_____	_____
$n = 25$ $S = 56$ $\alpha = 0.01$	_____	_____
$n = 35$ $S = 0.15$ $\alpha = 0.10$	_____	_____
$n = 78$ $S = 1.56$ $\alpha = 0.05$	_____	_____
$n = 95$ $S = 8.9$ $\alpha = 0.05$	_____	_____

26. Lead poisoning is a common problem for young children who live in older houses where lead-based paint was used and is now peeling and flaking from walls and ceilings. The paint has a sweet taste, and children consume it both purposefully and inadvertently as paint dust on objects they put in their mouth. Lead is not excreted by the kidneys and accumulates in the body, where it can cause neurological problems including brain damage. A study was done to assess the effectiveness of a new drug designed to mobilize lead in a form that will be excreted, thereby reducing blood lead levels. A sample of 59 children less than 6 years old who were being treated for lead poisoning was measured for blood lead concentration (µg/dl), given the new drug for a period of 28 days, and then blood lead levels were remeasured. Summary statistics for these data are given below:

	n	Mean	St. Dev.	Min.	Max.
Pre-test	59	40	6	25	66
Post-test	59	23	8	11	78
Difference (Post−Pre)	59	−17	9	−36	+25

A boxplot of the difference values indicates that the data distribution was moderately positively skewed, with two positive outliers corresponding to children who had been re-exposed to lead paint during the treatment period. Do these data provide sufficient evidence for the company to claim that their new drug is effective for lowering blood lead levels in children suffering from lead poisoning? *Note:* The investigators are only interested in a difference that shows the drug lowers lead concentration. Hence, you should perform a one-tailed test.

a. Based on the information provided above, assess whether or not the Normality assumption for the paired *t*-test is valid. Explain your answer.

b. State the Null and Alternative hypotheses.

H$_0$: _____   H$_a$: _____

c. Using the information in the question, scale the X-axis for the sampling distribution $\overline{D}$ under the assumption that the Null hypothesis is true. Shade the area under the curve that corresponds to the *p*-value.

$E(\overline{D})$ = _____

$S_{\overline{D}}$ = _____

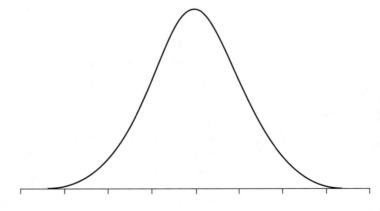

    d. Compute the *t*-test statistic and use the *t*-table to determine the *p*-value.

    e. Interpret the meaning of this *p*-value from the paired *t*-test in the context of this study. Use terms understandable to someone who has not studied statistics.

    f. Based on this *p*-value, state the appropriate conclusion in the context of this study. Use a format appropriate for a scientific journal.

    g. Having stated your conclusion, explain the second interpretation of the *p*-value.

    h. Compute the 95% confidence interval for the mean difference in blood lead concentration associated with this new drug treatment.

    i. Explain the interpretation of this confidence interval in the context of this study. Use words understandable to someone who has not studied statistics.

27. Suppose that the Food and Drug Administration stipulates that the drug in Problem 26 must reduce blood lead levels by 20 µg/dl to be considered effective. Determine the power of this study to detect an average decrease in blood lead concentration of $\mu_D = -20$, using a one-tailed test with $\alpha = 0.05$.

a. Scale the X-axis of the sampling distribution of $\overline{D}$ under the assumption of the Null hypothesis. Shade the area under the curve that corresponds to the probability of obtaining average difference values that would provide sufficient evidence to claim the drug is effective.

$E(\overline{D}) =$ _____

$S_{\overline{D}} =$ _____

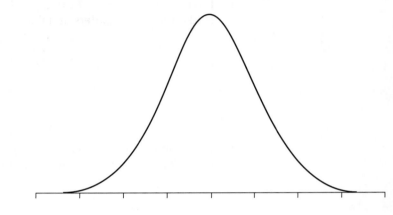

b. Compute the smallest value for $\overline{D}^{*}$ that would provide sufficient evidence to reject the Null hypothesis with a *p*-value $= 0.05$.

c. Scale the X-axis of the sampling distribution of $\overline{D}$ if the new drug actually reduces blood lead concentration by 20 µg/dl. Shade the area under this curve that corresponds to the power of the test.

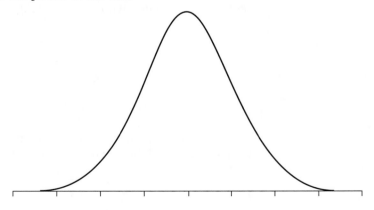

d. Determine the power $P[\overline{D} \leq \overline{D}^*$ if $\mu_D = -20]$.

e. Explain the meaning of this power value in this context. Use terms understandable to someone who has not studied statistics.

28. Suppose that the study described in Problem 26 had been done as a completely randomized experiment, and the summary statistics presented for the pre-test and post-test data were for the control and treatment group instead.

	n	Mean	St. Dev.	Min.	Max.
Control Group	59	40	6	25	66
Treatment Group	59	23	8	11	78

Suppose that boxplots of the blood lead concentration for the two groups are both moderately positively skewed, and the treatment group data include two large outliers corresponding to children who had been re-exposed to lead paint during the treatment period. Perform the following analyses to determine if these data provide sufficient evidence for the company to claim that their new drug is effective for lowering blood lead levels in children suffering from lead poisoning.

a. Based on the information provided above, assess whether or not the Normality assumption for the two-sample t-test is valid. Explain your answer.

b. Perform the F_{max} test and assess whether or not the equal-variances assumption for the pooled-variance t-test is fulfilled.

c. Based on the results of the F_{max} test, compute the standard error of the difference between the two sample means $S_{(\bar{x}_C - \bar{x}_T)}$.

d. State the Null and Alternative hypotheses.

H_0: _____ H_a: _____

e. Scale the X-axis for the sampling distribution $(\bar{x}_C - \bar{x}_T)$ under the assumption that the Null hypothesis is true. Shade the area under the curve that corresponds to the p-value for a one-tailed test.

$E(\bar{x}_C - \bar{x}_T)$ = _____

$S_{(\bar{x}_C - \bar{x}_T)}$ = _____

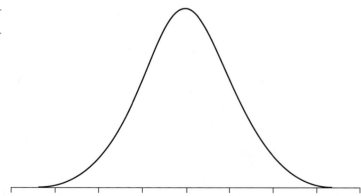

f. Compute the *t*-test statistic and use the *t*-table to determine the *p*-value.

g. Interpret the meaning of this *p*-value from the two-sample *t*-test in the context of this study. Use terms understandable to someone who has not studied statistics.

h. Based on this *p*-value, state the appropriate conclusion in the context of this study.

i. Having stated your conclusion, explain the second interpretation of the *p*-value.

j. Compute the 95% confidence interval for the difference in mean blood lead concentration between the control and treatment groups.

k. Explain the interpretation of this confidence interval. Use words understandable to someone who has not studied statistics.

29. Suppose that the Food and Drug Administration stipulates that for the study in Problem 28 the drug must reduce blood lead levels by 20 µg/dl to be considered effective.

Perform the following analyses to determine the power of this study to detect a decrease in blood lead concentration of $(\mu_T - \mu_C) = -20$.

a. Scale the X-axis of the sampling distribution of $(\bar{x}_C - \bar{x}_T)$ under the assumption of the Null hypothesis. Shade the area under the curve that corresponds to the probability associated with values for $(\bar{x}_C - \bar{x}_T)$ that would provide sufficient evidence to reject the Null hypothesis, based on a one-tailed test with $\alpha = 0.05$.

$$E(\bar{x}_C - \bar{x}_T) = \underline{\hspace{1cm}}$$

$$S_{(\bar{x}C - \bar{x}T)} = \underline{\hspace{1cm}}$$

b. Compute the smallest value for $(\bar{x}_C - \bar{x}_T)^*$ that would provide sufficient evidence to reject the Null hypothesis with a p-value ≤ 0.05.

c. Scale the X-axis of the sampling distribution of $(\bar{x}_C - \bar{x}_T)$ if the new drug actually reduces mean blood lead concentration by 20 µg/dl. Shade the area under this curve that corresponds to the probability $P[(\bar{x}_C - \bar{x}_T) \geq (\bar{x}_C - \bar{x}_T)^*$ if $(\mu_T - \mu_C) = -20]$.

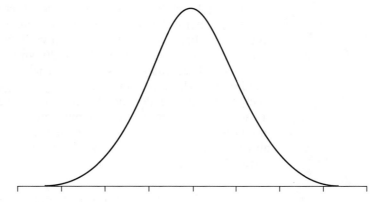

d. Determine the power of this study.

e. Explain the meaning of this power value. Use terms understandable to someone who has not studied statistics.

Chapter 9 Study Problems

1. It is well documented in public health literature that a greater proportion of African American ("black") women are obese as compared to non-Hispanic, Caucasian ("white") American women. One hypothesis for this difference is that black women have a lower basal metabolic rate than white women. To test this hypothesis, the following study was done: 20 black and 20 white obese women were randomly selected from hospital records and asked to participate in the study. Each black woman was paired with a white woman of similar age, weight, and height, because these variables affect metabolic rate. Resting metabolic rate (kilojoules per day or kJ/d) was measured for each woman. The statistics below were computed for the groups of black and white women separately, and for the difference, D = White − Black, between each matched-pair of women. Graphical analyses of data for each group and the difference values all indicated a Normal distribution.

	Group Statistics			Matched-Pairs Statistics		
	n	$\bar{x}$	S	n_D	$\bar{D}$	S_D
Black women	20	6056	911	20	550	950
White women	20	6606	984			

   a. *Explain why* the paired t-test is the most appropriate for analyzing these data.

   b. State the appropriate Null and Alternative hypotheses for this matched-pairs study.

   H_0: _____ H_a: _____

   c. Compute the standard error of $\bar{D}$ ($S_{\bar{D}}$).

d. Scale the X-axis of the sampling distribution of $\overline{D}$ under the premise that H_0 is true, and shade the area under the curve that corresponds to the *p*-value.

$E(\overline{D})$ = _____

$S_{\overline{D}}$ = _____

Shape: _____

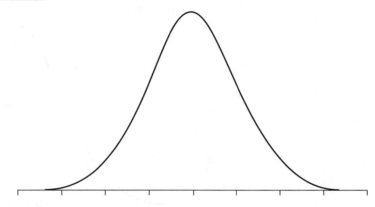

e. Compute the *t*-test statistic and determine the degrees of freedom and *p*-value.

f. State your conclusion with regard to the original question.

2. Compute the power of the paired *t*-test in Study Problem 1 to detect a 500 kJ/d mean difference in resting metabolic rate between black and white obese women, with α = 0.05.

 a. Scale the X-axis of the sampling distribution of $\overline{D}$ under the Null hypothesis. Shade the area under the curve that corresponds to the probability associated with values of $\overline{D}$ that would provide sufficient evidence to reject H_0 using a *two-tailed test* with a Type I error rate of α = 0.05.

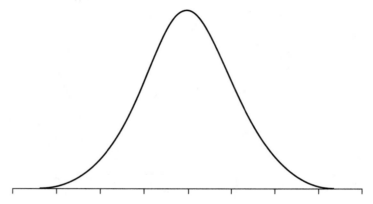

 b. Determine the minimum significant mean difference $\overline{D}^*$ at the p = 0.05 significance level.

 c. Scale the X-axis of the sampling distribution of $\overline{D}$ under the premise that the true difference between white and black women was 500 kJ/d. Shade the area under the curve that corresponds to the probability associated with values of $\overline{D}$ that would provide sufficient evidence to reject H_0 with a Type I error rate of α = 0.05, based on a two-tailed test.

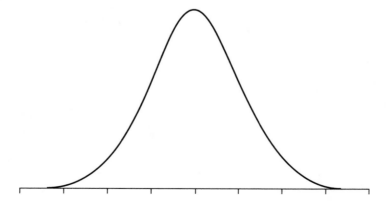

d. Determine the power of this study $= P[\overline{D} \geq \overline{D}^*$ if $\mu_D = 500$ kJ/d].

3. Suppose that the study described in Study Problem 1 had been done as a completely randomized design with two independent groups. Use the summary statistics for the two groups presented in Study Problem 1 to perform the appropriate two-sample t-test for this study.

    a. Explain whether or not the information provided indicates the assumptions for the two-sample t-test are fulfilled.

    b. Compute the standard error of the differences between the sample means, $S_{(\bar{x}_w - \bar{x}_b)}$.

    c. Scale the X-axis of the sampling distribution under the assumption that the Null hypothesis is true, and shade the area under the curve that corresponds to the p-value of the two-sample t-test.

        $E(\bar{x}_w - \bar{x}_b) = $ _____

          $S_{(\bar{x}_w - \bar{x}_b)} = $ _____

        Shape: _____

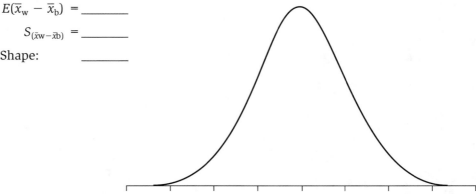

d. Compute the *t*-test statistic and determine the df and the *p*-value.

e. Explain what the *p*-value represents in this context. Use terms that a nonstatistician could understand.

f. State your conclusion with regard to the original scientific question.

4. Compute the power of this two-sample t-test to detect a 500 kJ/d difference between the mean for obese black women and the mean for obese white women, with a two-tailed Type I error rate $\alpha = 0.05$.

a. Scale the X-axis of the sampling distribution of $(\bar{x}_w - \bar{x}_b)$ under the premise that the Null hypothesis is true. Shade the area under the curve that corresponds to values of $(\bar{x}_w - \bar{x}_b)$ that would provide sufficient evidence to reject H_0.

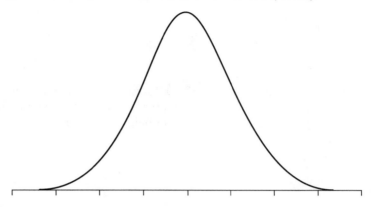

b. Compute the smallest significant difference $(\bar{x}_w - \bar{x}_b)^*$.

c. Scale the X-axis of the sampling distribution of $(\bar{x}_w - \bar{x}_b)$ under the premise that $(\mu_w - \mu_b) = 500$. Shade the area under the curve that corresponds to values of $(\bar{x}_w - \bar{x}_b)$ that would provide sufficient evidence to reject H_0, based on a two-tailed, two-sample t-test with $\alpha = 0.05$.

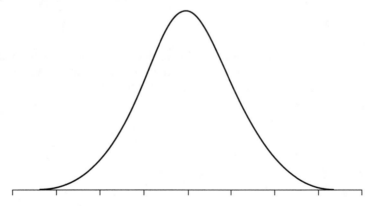

d. Compute the power $= P[(\bar{x}_w - \bar{x}_b) \geq (\bar{x}_w - \bar{x}_b)^*$ if $(\mu_w - \mu_b) = 500]$.

    e. Compare the power of the two-sample *t*-test computed here to the power of the paired *t*-test computed in Study Problem 2. Which of these two types of *t*-test is more powerful? *Explain why* this test is more powerful.

5. a. Compute the 95% confidence interval for the true mean difference μ_D in metabolic rate between black and white women who were matched based on similarity of age, weight, and height. Use summary statistics presented in Study Problem 1.

    b. Explain the interpretation of this confidence interval in this context. Use terms understandable to someone who has not studied statistics.

    c. Compute the 95% confidence interval for the true difference in mean basal metabolic rate between black and white women ($\mu_w - \mu_b$). Use summary statistics presented in Study Problem 1.

    d. Explain the interpretation of this confidence interval in the context of this study. Use terms understandable to someone who has not studied statistics.

    e. Compare the margin of error for the confidence interval for μ_D with the margin of error for the confidence interval for $(\mu_w - \mu_b)$. Explain why the margin of error for one is much smaller than that of the other.

Name _____   Date _____

Answer Key to Chapter 9 Study Problems

1. a. When a matched-pairs study design is used to obtain the data, the paired *t*-test is the most powerful way to analyze the data.

b. H_0: $\mu_D = 0$ H_a: $\mu_D \neq 0$,   where: D = White − Black.

c. $S_{\bar{D}} = S_D / \sqrt{n} = 950 / \sqrt{20}$

   $= 212.4$

d. $E(\bar{D})$ $= 0$

   $S_{\bar{D}}$ $= 212.4$

   Shape: **Normal**

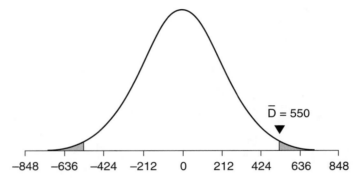

e. $P[\bar{D} \geq 550$ if $\mu_D = 0] = P[t \geq (550 - 0) / 212.4]$

   $= P[t \geq 2.589]$ with df $= (n - 1) = 19$

   $p < 0.02$ ($= p < 0.01 \times$ two-tails)

f. *Conclusion*: Black women have lower basal metabolic rate than white women of similar age, weight, and height ($p < 0.02$).

2. a.

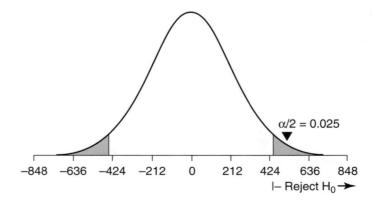

b. $\overline{D}^* = t_{0.025,\text{df}=19} \, S_{\overline{D}}$

$\qquad = 2.093 \ (212.4)$

$\qquad = \mathbf{444.6}$

c.

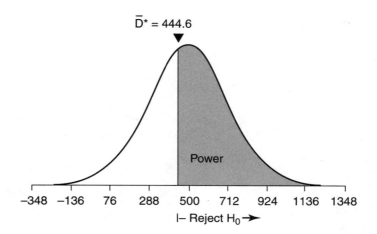

$\overline{D}^* = 444.6$

Power

−348 −136 76 288 500 712 924 1136 1348

⊢ Reject H₀ ➝

d. Power $= P[\overline{D} \geq 444.6 \text{ if } \mu_D = 500] = P[t \geq (444.6 - 500) / 212.4]$

$\qquad = P[t \geq -0.261]$

0.50 < Power < 0.75

3. a. Assumptions:

Randomized, unbiased study design: It is given that investigators performed a randomized two-sample study.

Sampling distribution of the difference between means is Normal: It is given that the data distributions for the two groups were approximately Normal, with no outliers. The Central Limit Theorem states that with $n = 20$ in each group, the sampling distribution of each group mean can be assumed Normal. Hence, the sampling distribution of $(\bar{x}_w - \bar{x}_b)$ can be assumed Normal.

Equal variances (for pooled-variance t-test): Based on F_{max} test below, sample variances are sufficiently similar to assume that the population variances are equal.

$F_{max} = 984^2 / 911^2 = 1.17$

$F_{critical}$ with df $= 19 = 2.86$ (df rounded down to 15)

b. Compute $S_{(\bar{x}w - \bar{x}b)}$.

$S_p = \sqrt{(S^2_w + S^2_b) / 2} = \sqrt{(984^2 + 911^{\,2)}) / 2} = \sqrt{899088.5} = 948.2$

$S_{(\bar{x}w - \bar{x}b)} = S_p \sqrt{(1/n_w + 1/n_b)} = 948.2 \sqrt{(1/20 + 1/20)} = \mathbf{299.8}$

c. $E(\bar{x}_w - \bar{x}_b) = 0$

$S_{(\bar{x}w-\bar{x}b)} = 299.8$

Shape: **Normal**

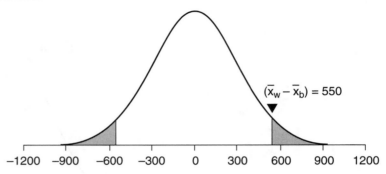

$(\bar{x}_w - \bar{x}_b) = 550$

d. $P[(\bar{x}_w - \bar{x}_b) \geq 550 \text{ if } (\mu_w - \mu_b) = 0] = P[t \geq (550 - 0) / 299.8]$

$= P[t \geq 1.835] \text{ with df} = n_w + n_b - 2 = \mathbf{38}$

$\boldsymbol{p} < \mathbf{0.10}$ (0.05 × two-tails)

e. This p-value is the probability that the observed difference between the mean basal metabolic rate for white and black women is due only to random variation (i.e., there is no difference between white and black women).

f. *Conclusion:* The results of this study suggest that black women have lower basal metabolic rate than white women, but more data are needed to verify this.

4. a.

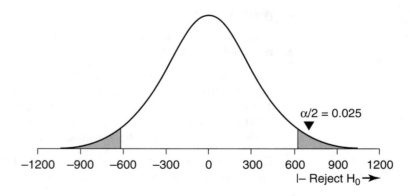

$\alpha/2 = 0.025$

|– Reject H_0 →

b. Minimum significant difference $(\bar{x}_w - \bar{x}_b)^* = t_{0.025, df = 38} \, S_{(\bar{x}w - \bar{x}b)}$

$= 2.042 \, (299.8)$

$= \mathbf{612.2}$

c.

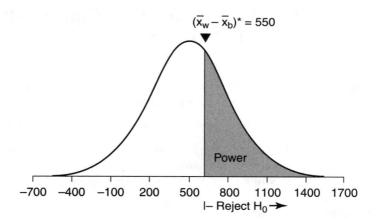

$(\bar{x}_w - \bar{x}_b)^* = 550$

Power

$-700 \quad -400 \quad -100 \quad 200 \quad \quad 500 \quad \quad 800 \quad 1100 \quad 1400 \quad 1700$

$\vdash$ Reject $H_0 \rightarrow$

d. Power $= P[(\bar{x}_w - \bar{x}_b) \geq 612.2 \text{ if } (\mu_w - \mu_b) = 500] = P[t \geq (612.2 - 500) / 299.8]$

$$= P[t \geq 0.374]$$

0.25 < Power < 0.50

e. The matched-pairs study design and paired t-test had approximately double the power of the completely randomized design and two-sample t-test.

Computing the difference values between individuals in matched-pairs reduces random variation, such that $S_{\bar{D}} < S_{(\bar{x}_w - \bar{x}b)}$. Reducing random variation always increases power.

5. a. 95% confidence interval for $\mu_D = \bar{D} \pm t_{0.025, df = 19} (S_{\bar{D}})$

$$= 550 \pm 2.093 \ (212.4)$$

$$= \textbf{550} \pm \textbf{444.6}$$

b. Because 95% of all confidence intervals computed by this method will include the true mean difference μ_D, we are confident that the true mean difference in basal metabolic rate between white and black women of similar age, height, and weight, falls within the range 550 ± 444.6.

c. The 95% confidence interval for $(\mu_w - \mu_b) = (\bar{x}_w - \bar{x}_b) \pm t_{0.025, df = 38} \ S_{(\bar{x}w - \bar{x}b)}$

$$= 550 \pm 2.042 \ (299.8)$$

$$= \textbf{550} \pm \textbf{612.2}$$

d. Because 95% of all confidence intervals computed by this method will include the true difference between two population means $(\mu_1 - \mu_2)$, we are 95% confident that the true difference in mean basal metabolic rate between obese white and black women falls within the range 550 ± 612.2.

e. The margin of error of the 95% confidence interval for μ_D is smaller than that for the 95% confidence interval for $(\mu_w - \mu_b)$ because the random variation among differences between matched-pairs of subjects $(S_{\bar{D}})$ is less than the random variation between means for two groups of randomly assigned subjects $S_{(\bar{x}w - \bar{x}b)}$.

Less random variation results in narrower confidence intervals, and more precise estimates of the true difference in basal metabolic rate between obese black and white women.

Exercise A: Tests of Significance for Means and Medians

Objective

1. You will learn how to use descriptive summary statistics, boxplots, and Normal quantile plots to verify that data meet the assumptions for the two-sample t-test.

2. You will learn to transform data so that they better meet the Normality assumption.

3. You will compare results from separate- and pooled-variance t-tests and the nonparametric Mann-Whitney U-test and be able to describe the consequences of using tests when their assumptions are violated.

Introduction

The data used in this exercise were derived from a forestry "spacing trial." Red pine seedlings were planted with 4 ft between rows and adjacent plants in one plot, and with 8-ft spacing in another. The wider the spacing, the greater the availability of light, water, and nutrients for each tree. Hence, there is biological reason to expect that trees planted at wider spacing should grow faster and become larger than trees planted at close spacing. However, this increase in growth occurs only within the range of spacings that individual trees are capable of filling. At very wide spacing, trees are unable to grow large enough to utilize all the available space, and any additional increase in spacing provides no additional increase in tree growth. The question addressed by this study was, "Do red pine planted at 8-ft spacing grow faster than red pine planted at 4-ft spacing?" The data values are the total volume of wood (dm^3) in the stems of individual 55-yr-old trees randomly sampled from the two adjacent experimental plots.

RP4	102	125	130	230	206	271	95	159	191	63	150	296	117	157	121
RP8	272	111	134	615	230	200	80	314	76	141	355	470	501	113	61

Instructions

1. Enter the data into a computer statistics program as two variables, named **RP4** and **RP8**.

2. Perform graphical and statistical analyses on each variable to determine if the data fulfill the assumptions for two-sample t-tests (side-by-side boxplots; Normal quantile plots; summary statistics, including mean, median, standard deviation, minimum, maximum).

3. Create two new variables, named **LogRP4** and **LogRP8** that contain the log-transformed data values (transformation done to address non-Normality in the data). Perform graphical and statistical exploratory data analyses on the transformed data. See additional instructions for this in the computer tutorial titled Data Manipulation.

4. Perform the following tests of significance on these data.

    a. Appropriate two-sample t-test on the original data. Use an F_{max} test to determine which of the separate-variance or pooled-variance t-test is appropriate.

    b. Appropriate two-sample t-test on the log-transformed data. Again, apply an F_{max} test to the transformed data values to determine if the separate-variance or pooled-variance t-test is appropriate.

    c. Nonparametric Mann-Whitney U-test on the original data.

5. Copy/Paste the graphical and statistical analyses into a word processor, arranged on pages as described below:

Page 1: Include graphical and statistical exploratory data analyses of original data and results of two-sample *t*-test on these data.

Page 2: Include graphical and statistical exploratory data analyses of the log-transformed data and results of two-sample *t*-test on these data.

Page 3: Include results of the Mann-Whitney *U*-test.

Report   Answer the following questions *in complete sentences and paragraphs* that are understandable without referring back to the questions. *Your answers should be typed.*

1. List *all* assumptions for the two-sample *t*-test. For each assumption, state whether or not the original data in Vol4 and Vol8 meet that assumption. If not, describe how the assumption is violated *and explain how you determined the assumption was violated.* Your assessment of the Normality assumption must take into account the sample size and the Central Limit Theorem. Refer to the boxplots and/or Normal probability plots and describe the manner in which one or both data distributions deviate from Normal.

2. Do the log-transformed data in LogRP4 and LogRP8 better meet the assumptions? Explain your answer as described in Question 1 above.

3. For each of the three tests of significance, state the appropriate conclusion in terms of the original scientific question. Your statement should include the *p*-value in parentheses, in a format similar to that used in scientific writing.

4. You have performed three different two-sample tests of significance during this exercise. For which test(s) of significance did the data meet all assumptions? *Among only those tests of significance that are valid,* which test maximized the statistical power of the analysis? (A smaller *p*-value indicates greater power.) Explain why this test has greater power.

5. An important decision that scientists make during data analysis is choosing the most appropriate statistical test so that they can use their data to draw valid conclusions about the question of interest. Based on specific comparisons between the outcomes of the three tests of significance you assessed in Question 4 above, describe the consequences of selecting statistical tests that are *not* appropriate.

Name _____   Date _____

4. Boxplots, Normal quantile plots, and sample sizes for data obtained by two studies, (a) and (b), are given below. For each study, state whether the *F*-test or the Levene test is more appropriate to test the hypothesis of equal population variances. Explain your answers.

(a) $n_1 = 20$ $n_2 = 20$

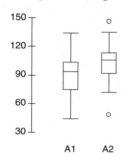
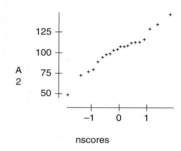

Which test is most appropriate? _____

Why? _____

_____

(b) $n_1 = 100$ $n_2 = 100$

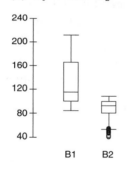
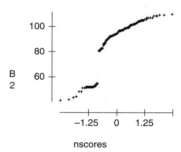

Which test is most appropriate? _____

Why? _____

_____

Name _____   Date _____

Study Problems for Chapter 10

1. For each of the following sets of two sample variances and sample sizes: (a) Compute the F_{test} statistic, (b) Determine the numerator and denominator degrees of freedom, (c) Obtain the critical F-value for $\alpha = 0.05$ from the F-distribution (Appendix Table 5), (d) State your conclusion with regard to whether or not the sample data provide sufficient evidence to claim that the population variances are not equal, and (e) Use a computer spreadsheet program to obtain the exact p-value for the computed F_{test} value.

S_1^2	S_2^2	n_1	n_2	F_{test}	DFN	DFD	$F_{critical}$	Sig? (Y/N)	p-Value
65	41	20	28	____	____	____	____	____	____
1.287	2.349	45	45	____	____	____	____	____	____
591	2687	15	37	____	____	____	____	____	____

2. A researcher has performed a completely randomized experiment on the efficacy of a new drug for stimulating weight loss in clinically obese people. Initial exploratory data analysis indicates that the distributions of 25 data values from each of the two experimental groups are approximately Normal. She will use a two-sample t-test to compare the means for the control and treatment group, and she must determine which of the pooled-variance or separate-variance t-test is most appropriate. The *standard deviation* for the control group data values is $S_c = 12.9$, and the standard deviation for the treatment group data values is $S_t = 7.4$. Perform an F-test to determine if the equal variances assumption of the pooled-variance t-test is fulfilled.

   a. Compute the F_{test} statistic.

   b. Enter the numerator and denominator degrees of freedom:

   DFN = _____   DFD = _____

   c. Obtain the critical F-value from the F-distribution Appendix Table 5: _____

   d. State your conclusion regarding whether or not it would be valid for the investigator to use the pooled-variance t-test. Explain your answer.

3. Each year newspapers report results from the Scholastic Aptitude Test (SAT) for different schools within a local region, and compare results among different regions and states. Many administrators of schools that have below average SAT scores complain this comparison is unfair. In some schools, only those students with definite intentions of going on to college take the SAT. That is, only the most academically successful students take the test. In other schools, the administration encourages all students to take the SAT and a much larger percentage of the student body take the test. Administrators in these schools believe that participation in the SAT by a wider diversity of students results in lower average scores for their schools. If this were the reason for lower average SAT scores, the other consequence should be that the variation of SAT scores among students at schools that encourage wider participation should be greater than at schools where only college-bound students take the test. If a school has below average SAT scores, but the variance is no greater than a school with higher scores, one could conclude that the first school does not prepare its students as well as the second. A newspaper reports the following summary statistics for SAT scores obtained by students at two local schools.

School A: Mean Verbal SAT: 501   Standard deviation: 113 $n = 215$

School B: Mean Verbal SAT: 565   Standard deviation: 72 $n = 97$

School A encourages all students to take the SAT. Only college-bound students take the SAT at School B. Suppose that the data distributions for both schools are Normal.

Perform an *F*-test to determine if the variance of SAT scores is greater at the school that encourages all students to participate in the SAT. Show all work. State your conclusion in the context of the original question.

Answer Key for Chapter 10 Study Problems

1.

S_1^2	S_2^2	n_1	n_2	F_{test}	DFN	DFD	$F_{critical}$	Sig? (Y/N)	p-Value
65	41	20	28	1.59	19	27	1.99	N	0.132
1.287	2.349	45	45	1.83	44	44	1.69	Y	0.024
591	2687	15	37	4.55	36	14	2.28	Y	0.002

2. a. $F_{test} = 12.9^2 / 7.4^2 = 166.41 / 54.76 = \mathbf{3.04}$

b. DFN = **24** DFD = **24**

c. Critical F-value from Appendix Table 5: **1.98**

d. *Conclusion:* The researcher should use the separate-variance *t*-test. The *F*-value computed from the ratio of the two sample variances is much larger than the critical *F*-value for $\alpha = 0.05$. This provides strong evidence that the variances of the two populations are not equal.

3. $F_{test} = 113^2 / 72^2 = 12769 / 5184 = \mathbf{2.46}$

DFN = **214**   DFD = **96**   *F*-critical = **1.36**

Conclusion: The variance of SAT scores at School A is significantly greater than the variance of scores at School B ($p < 0.05$). This indicates that the policy at School A of encouraging all students to take the SAT results in a wider diversity of students being included. This may at least partially explain the lower average SAT score at this school.

Name _____ Date _____

Exercise A: Performing the Levene Test Using MS Excel

Objective   You will learn how use MS Excel to perform the calculations for the Levene test for equality of variances, including (1) entering data, (2) entering formulae, (3) using functions in formulae (mean, absolute value), and (4) using statistical functions (two-sample t-test).

Introduction   In the text, I described a study that compared the effectiveness of highly active antiretroviral drug therapy (HAART) and intermittent drug therapy (IDT) for controlling HIV blood viral loads. HAART requires a complex daily drug regime that is very expensive and has many adverse side effects. IDT involves the same drug cocktail as HAART, but the patients take the drugs in a cycle of 7 days "on" and 7 days "off" the therapy. This IDT regime cuts both the cost and adverse side effects. The purpose of the study was to determine if IDT provided the same control over HIV blood viral loads. In Example 9.13 of the textbook, a Mann-Whitney U-test was performed to determine if median blood viral load differed between these two therapies. Another possible difference between the two therapies is that patients may find it more difficult to track the intermittent regime than a regime that requires they follow the same routine every day. This could lead to more variable results in patients, such that those who fail to keep track of the IDT schedule may have substantially higher viral loads than other patients who scrupulously follow the IDT regime. The investigator wants to determine if the variance of blood viral load differs between the two treatment groups.

Data

HAART	310	102	114	61	96	207	137	78	85	23
IDT	127	114	499	51	28	348	205	67	70	111

Instructions   1. Enter the data for these two groups of experimental subjects in two columns of an MS Excel spreadsheet, one with the column title **HAART** and the other **IDT**. The column titles should be in spreadsheet cells A1 and B1, and the data values should be in cells A2–A11 and B2–B11.

2. *Compute the sample means:* In cell A12, enter the formula to compute the mean of the data values for the HAART group.

   a. Put the mouse arrow in cell **A12** and click the mouse button to highlight that cell.

   b. Near the top of the screen, just above the letters that label the spreadsheet columns is a white area (line) with a = symbol in front. This is the "formula bar" where you enter formulae to perform calculations. You must first click on the = symbol to tell Excel that what you are about to type is a calculation formula and not simple text.

   c. After clicking on = , type the following *exactly as it is shown* into the white area of the formula bar: **Average(A2:A11)**. After typing the formula, press the Return key. The average of the 10 data values for the HAART group should appear.

   d. To compute the average for the IDT group, copy the formula from cell A12 to cell B12. Click on cell **A12**. A black border should appear. In the lower-right corner of this border is a small box. Place the point of the mouse arrow on this box and it should turn into a black + . Press and hold down the left mouse button, move the mouse arrow over to cell **B12**, and then release the mouse button. The mean for the IDT data values should appear.

3. *Compute the absolute deviation values:* The Levene test is based on a comparison of the mean absolute deviation (AD) between data values and their group mean. To perform the test you must compute the individual AD values.

   a. In cell C1 type the title **AD-H**, and in D1 type **AD-I**.

   b. Click on cell **C2**, click on the = symbol in front of the formula bar, and enter the following formula: **ABS(A2-A$12)**. ABS is the absolute value function in Excel, which strips off any negative signs from the differences between individual values and their mean. The $ in the formula "locks" the "12" part of the cell address A12. When this formula is copied to other cells, this number will not change.

   c. Copy the formula in cell C2 to cell D2 using the procedure described above.

   d. At this point both cells C2 and D2 should have the black border around them and the small black box should be in the lower-right corner of cell D2. Put the mouse arrow on this black box (should become a +), press and hold the left mouse button, drag the mouse arrow down to cell **D11**, and release the mouse button. AD values for all 10 data values in both groups should appear.

4. *Perform a two-sample t-test on the AD-values* to determine if there is sufficient evidence to conclude that variance of HIV blood viral loads differs between HAART and IDT groups.

   a. Click on cell **D12**, click on the = symbol in front of the formula bar, and enter the following text in the formula bar exactly as shown: **TTEST(** A "dialog box" should appear that guides you through how to enter the remaining specifications for the *t*-test.

   b. In the first white field, labeled Array 1, enter **C2:C11**.

   c. In the second white field, labeled Array 2, enter **D2:D11**.

   d. In the third white field, labeled Tails, enter **2**. (This should be a two-tailed test.)

   e. In the fourth white field, labeled Type, enter **3** (specifying you want a separate-variance two-sample *t*-test).

   f. When finished, click on **OK** and the *p*-value for the *t*-test should appear in cell D12. In cell E12 type the text **Levene test p-value** to be a label for this result.

5. *Perform the F-test for equal variances.*

   a. *Compute the variances for HAART and IDT data values.* Using the procedure for entering a formula described above, enter the following formula into cell A13: **VAR(A2:A11)**. The variance for the HAART data should appear.

   b. Using the procedure described above, copy this formula to cell B13.

   c. *Compute the F_{test}.* In cell A14 type the text **F-test**. In cell B14, use the procedure for entering formulae described above to enter whichever one of the following two formulae results in dividing the larger of the two variances by the smaller: **A13/B13** or **B13/A13**.

   d. *Determine the p-value for the F-test.* In cell A15, type the text **p-value**. In cell B15, enter the following formula: **FDIST(B14,9,9)**. This stipulates that the *F*-test statistic is in cell B14, the degrees of freedom for both the numerator variance and the denominator variance are 9. The *p*-value for the *F*-test should appear.

6. *Print the spreadsheet.*

    a. Drag the mouse arrow across the spreadsheet from cell A1 to cell E15. The entire block of data and calculations should be highlighted in black.

    b. In the menu bar at the top of the screen, select **Edit/Copy**.

    c. Click on the **Start** button in the lower-left corner of the screen, select **Programs** from the pop-up menu, and double-click on the icon for **Microsoft Word**. This will open a new Word document on the screen, covering the Excel spreadsheet.

    d. At the top of the MS Word page type the title **Chapter 10 Computer Exercise A**. Below that type your name, and then press the Enter key twice to move the cursor down the page.

    e. In the menu bar at the top of the MS Word screen, select **Edit/Paste**. The data and calculations copied from the Excel spreadsheet should appear on the MS Word page.

Report   *Type answers* to the following questions on the MS Word page immediately below the spreadsheet data and calculations.

    1. Look at the exploratory analyses of these data in Example 9.13 of the textbook. Describe how these data violate the assumptions for the F-test of equality of variances.

    2. *Describe why* the Levene test is more appropriate for comparing the variances of these two samples than the F-test. That is, state why the Levene test is less sensitive to violation of the Normality assumption.

    3. Compare the p-values from the Levene test and the F-test. Based on this comparison, describe the consequence of applying the F-test to these data when the assumptions for this test are not fulfilled.

2. A researcher is testing the effectiveness of the Atkins (high protein, low carbohydrate) diet for improving health. This diet plan is purported to reduce weight, reduce blood sugar, reduce high blood pressure, reduce "bad" (LDL) cholesterol, and increase "good" (HDL) cholesterol. The research involves a Before-After study with 12 subjects. Each subject is measured for these five variables before starting the diet and again after following the diet for 12 weeks. Separate paired t-tests are applied to the data for each of these variables to determine if there is evidence that the diet had an effect on any of the above-mentioned variables. The results of these t-tests are presented below. To test the global Null hypothesis that the Atkins diet had no effect, the investigator believes he should control experiment-wise Type I error rate to $\alpha = 0.05$ by using the sequential Bonferroni procedure. Perform this procedure: (a) Put the t-test results in the proper order in the table below, (b) compute the adjusted critical p-value for each of the t-tests, and (c) state your conclusion for each test and with regard to the overall Null hypothesis.

**************** p-Values from paired t-tests. ****************

Weight	Blood Sugar	Blood Pressure	LDL Cholesterol	HDL Cholesterol
$p = 0.03$	$p = 0.001$	$p = 0.045$	$p = 0.075$	$p = 0.01$

Variable	p-Value	Ntests	d	Critical p	Conclusion
_____	_____	____	____	_____	_____
_____	_____	____	____	_____	_____
_____	_____	____	____	_____	_____
_____	_____	____	____	_____	_____
_____	_____	____	____	_____	_____

3. For the following situations, identify the response variable and the treatment/group variable; then state the number of groups (k), the number of observations per group (n_i), and the overall sample size (N).

a. In a study to compare yield for three varieties of corn, seed for each variety was randomly assigned to four one-acre plots (12 plots total). The yield of corn from each plot (bushels per plot) was measured.

Treatment variable: _____ N _____ k _____ n_i _____

Response variable: _____

b. A microbiologist wants to compare the toxicity of five disinfecting cleaning products on a bacteria that is a common cause of food poisoning. She applies each cleaning product to 10 petri plates that contain the bacteria and measures the number of bacteria colonies present after 24 hours.

Treatment variable: _____ N _____ k _____ n_i _____

Response variable: _____

4. A pharmaceutical company is testing the efficacy of a new drug to lower cholesterol. An experiment is done with five experimental groups, a control (0 mg of drug in the placebo pill), and four groups that receive different doses of the drug (10, 20, 30, 40 mg of drug). If mean cholesterol of any one of the treatment groups is lower than for the control group, this would be considered evidence that the drug is effective.

  a. Describe why it would be inappropriate to use two-sample t-tests among all possible pairs of groups to determine if mean cholesterol differed among the treatment groups. Use the terms "comparison-wise error rate" and "experiment-wise error rate."

  b. With this study design, the investigator could use the Bonferroni procedure to adjust comparison-wise error rates for the multiple two-sample t-tests needed to compare all groups *or* the investigator could use ANOVA to test the overall Null hypothesis that the drug was not effective at any dose. Which approach is more appropriate? Explain.

Name _____   Date _____

6. Some researchers have proposed that mixtures of different plant species are more productive than a single species at similar plant density. They believe that differences in leaf and root morphology allow mixtures of species to more fully utilize available light, water, and nutrients than can a single species. To test this hypothesis, a student selected three plant species that differ in rooting depth and whether they grow upright or flat. She set up 10 pots for each of the following "treatments": (a) 18 seeds of species B, (b) 18 seeds of species C, (c) 18 seeds of species D, and (d) 6 seeds each of species B, C, D combined (for a total of 18 seeds). The plants were allowed to grow for 16 weeks, then all the plant biomass (roots and shoots) in each pot was harvested, dried to constant weight, and total plant biomass (g) per pot was determined. The resulting data are provided below.

Pot	Species B	Species C	Species D	All Species
1	57	72	66	71
2	57	52	45	75
3	65	58	69	62
4	49	48	60	71
5	50	54	68	65
6	56	64	64	76
7	61	57	61	63
8	62	57	54	55
9	46	68	71	72
10	36	61	64	66

Enter these data into a computer statistics program. *Note:* For ANOVA, some programs require that the data for the response variable be in a single column and that a second variable be created wherein you indicate to which treatment group each data value belongs. Create two variables, named **Group** and **Biomass**. In the group variable indicate group identity using the letters B, C, D, and ALL. You must have one of these classes of the categorical variable Group indicated for every data value in the **Biomass** variable.

Produce the following statistical print-outs and copy/paste them into a single word processor page: summary statistics (by Group), side-by-side boxplots that display the distribution of biomass for all four groups, and an ANOVA table. On the word processor page, perform the F_{max} test for equality of variances. Append that results page to this page of the workbook.

a. What is the treatment/group variable? What is the response variable?

Treatment: _____

Response: _____

b. Do the data from this experiment meet the assumptions for ANOVA? Describe how you determined whether or not each assumption was valid.

c. State the H_0 and H_a. Use subscripts to denote specific group means.

H_0: _____

H_a: _____

d. Explain the interpretation of the ANOVA p-value in the context of the original question. Use terms understandable to a nonstatistician.

e. Based on the results from ANOVA, what is the appropriate conclusion?

f. Perform Tukey's HSD procedure to determine which of the group means are significantly different from each other at the $p \leq 0.05$ level. List the group means in rank order, and use letter suffixes to indicate which means are significantly different at the $p \leq 0.05$ level.

What do you conclude from this test?

g. Does this study provide strong evidence that mixtures of plant species are more productive than single species planted at the same density? What is the scope of the inference that you can draw from this study and how does this inference relate to the overall scientific question?

h. Perform a planned contrast for the alternative hypothesis that productivity of the mixed species pots is greater than means for the single species pots ($\mu_{all} > \mu_B = \mu_C = \mu_D$).

7. Suppose you decided that the data from the experiment described in Question 6 violated the Normality assumption. Use a computer statistics program to perform a Kruskal-Wallis test on these data.

If your program does not offer the K-W test: Most spreadsheet and statistics programs provide a Rank function or menu item that will automatically create a new variable that contains the ranks of data values. Use this to create a new variable named **Rank:Biomass**. Select this new rank variable as the response variable (Y) and the Group variable as the treatment (X). Perform Analysis of Variance on the rank variable (equivalent to performing a Kruskal-Wallis test). You should also produce summary statistics by Group for this rank variable to obtain the mean ranks statistics needed for the multiple comparisons test for medians. Copy/Paste the statistical print-outs to a word processor page and append that page to this one. Answer the following questions.

a. State the H_0 and H_a.

H_0: _____

H_a: _____

b. Explain the interpretation of the p-value in the context of the original question. Use terms understandable to a nonstatistician.

c. Based on the results from this Kruskal-Wallis test, what is the appropriate conclusion?

d. Perform the multiple comparisons test to determine which of the groups are significantly different from each other at the $p \leq 0.05$ level. List the group medians (from Summary Reports) in rank order, and use letter suffixes to indicate which are significantly different at the $p \leq 0.05$ level.

What do you conclude from this multiple comparisons test?

Enter these data into a computer statistics program. Some programs require that the data for the response variable be in a single column, and that a second variable be created wherein you indicate to which treatment group each data value belongs. Create two variables, named **Group** and **LDL**. For the group variable, indicate group identity using the letter designations for groups in the boldface column titles in the data table above. You must have one of these classes of the categorical variable Group indicated for every data value in the LDL variable.

Produce the following statistical print-outs and copy/paste them into a single word processor page: (1) summary statistics (by Group), (2) side-by-side boxplots of LDL for the six groups in a single graph, and (3) ANOVA table. On the word processor page perform the F_{max} test for equality of variances and *type your conclusion* regarding this assumption. Append that page to this one.

a. What is the treatment/group variable? What is the response variable?

Treatment: _____

Response: _____

b. Do the data from this experiment meet the assumptions for ANOVA? Describe how you determined whether or not each assumption was valid, with specific references to summary statistics, graphics, and so forth.

c. State the H_0 and H_a. Use subscripts to denote specific group means.

H_0: _____

H_a: _____

d. Explain the interpretation of the ANOVA *p*-value in the context of the original question. Use terms understandable to a nonstatistician.

e. Based on the results from ANOVA, what is the appropriate conclusion?

f. Perform Tukey's HSD procedure to determine which of the group means are significantly different from each other at the $p \leq 0.05$ level. List the group means in rank order, and use letter suffixes to indicate which are significantly different at the $p \leq 0.05$ level.

What do you conclude from this test?

g. Does this study provide sufficient evidence to claim that a diet rich in saturated fat (B) or trans fatty acids (StM, Sh, SfM, SLM) results in higher blood LDL cholesterol than a diet low in these substances (SO)?

h. What is the scope of the inference that you can draw from this study?

i. Perform a planned contrast for the alternative hypothesis that using soybean oil as the primary source of dietary fat results in lower blood LDL cholesterol than all other dietary sources of fat included in this study. Show your calculations and state your conclusion.

10. Suppose you decided that the data from the experiment described in Problem 9 violated the Normality assumption. Use a computer statistics program to perform a Kruskal-Wallis test on these data.

If your program does not offer the Kruskal-Wallis test: Create a new variable named **Rank:LDL** that contains the ranks of the data values. Most statistics and spreadsheet programs have a function that will determine the ranks, usually in the format: **RANK** *< variable name >*. Select this rank variable as the response variable (Y) and the Group variable as the treatment (X). Perform an Analysis of Variance on the rank variable (equivalent to performing a Kruskal-Wallis test). You should also produce summary statistics by Group for the Rank:LDL variable to obtain the mean ranks statistics needed for the multiple comparisons test for medians. Copy/Paste the statistical print-outs to a word processor page and append that page to this one. Answer the following questions.

a. State the H_0 and H_a.

H_0: _____

H_a: _____

Name _____   Date _____

b. Explain the interpretation of the *p*-value in the context of the original question. Use terms understandable to a nonstatistician.

c. Based on the results from this Kruskal-Wallis test, what is the appropriate conclusion?

d. Perform the multiple comparisons test to determine which of the group medians are significantly different from each other at the $p \leq 0.05$ level. *Note:* Although this is a multiple comparisons test for medians, it is based on the *mean ranks* for the groups. List the group *medians* (from the summary statistics) in rank order, and use letter suffixes to indicate which are significantly different at the $p \leq 0.05$ level.

What do you conclude from this multiple comparisons test?

Name _____ Date _____

Chapter 11 Study Problems

1. A taxonomist is trying to determine if two populations of plants are sufficiently mor-
phologically different to be considered two different species. He collects random samples from both populations and obtains data for the flower characteristics listed below.
For each flower characteristic he performs a two-sample t-test. If the two means for any
one flower characteristic are significantly different, he will conclude that the two populations are different species.

Characteristic	Mean of Population 1	Mean of Population 2	p-Value
Petal length	3.4	3.0	0.01
Petal width	0.9	0.6	0.025
Sepal length	1.5	1.0	0.01
Sepal width	0.5	0.4	0.15
Anther length	1.8	1.2	0.001
Pistil length	0.9	1.1	0.10

a. If the taxonomist used the standard $\alpha = 0.05$ criterion for rejecting the Null hypothesis for each two-sample t-test, what would the comparison-wise and experiment-
wise error rates be? (*Hint: n* = 6 tests with **P** = 0.05 for Type I error for each test.)

Comparison-wise Type I error rate? _____

Experiment-wise Type I error rate ? _____

b. Use the sequential Bonferroni procedure to determine which of these differences between mean flower dimensions is significant while keeping the overall, experiment-
wise Type I error rate to 0.05 or less.

    c. Which flower dimensions, if any, differ significantly between these two populations?

    d. State your conclusion with regard to the original scientific question.

2. A scientist wants to study whether decomposition of organic matter is influenced by environmental conditions in the Arctic. Specifically, he wants to determine if temperature and water availability influence the activity of microbes that cause decomposition. The intensity of solar radiation on north-facing slopes is less than on south-facing slopes, with the result that north slopes have lower temperature. Because soil water drains downhill, water availability is greater on lower slope positions than on upper slope positions. The scientist expects decomposition to be fastest under warmer, moister conditions. To determine if variation in temperature and soil water actually does influence decomposition, he randomly assigns 11 samples of a uniform mix of dead leaf matter to each of the four site types: low-north slopes, upper-north slopes, low-south slopes, and upper-south slopes. He buries these samples in the top layer of soil for 10 weeks, and then measures the decrease in mass of the organic matter from each leaf sample (g/100g), presumably due to decomposition. Large values of mass loss indicate high decomposition rate. Summary statistics and ANOVA results are listed below.

Statistics for Mass Loss (grams)

	Low-South	Upper-South	Low-North	Upper-North
$\bar{x}$	7.92	5.38	4.48	5.01
Median	8.02	5.92	4.54	4.83
S	1.29	1.79	1.20	1.30

Analysis of Variance Table

Source	df	Sums of Squares	Mean Squares	F-Ratio	p-Value
Group	3	77.0	25.7	12.8	< 0.00001
Error	40	80.1	2.0		
Total	43	157.1			

a. What is the treatment variable? _____

b. What is the response variable? _____

c. Do these data meet the assumptions for ANOVA? Explain how you made your decision for each assumption.

d. State H_0 and H_a for the analysis of variance.

H_0: _____

H_a: _____

e. Describe the interpretation of the p-value in the ANOVA table in the context of this study question. Use terms understandable to nonstatisticians.

f. Based on the results presented in the ANOVA table above, what can you conclude with regard to the original question?

g. Perform a Tukey's HSD test to determine which of the four group means are significantly different at the $p = 0.05$ level.

State your conclusion with regard to which treatment groups are or are not different.

h. Perform a planned contrast for the Alternative hypothesis that mean decomposition rate should be higher on the low-south slopes (where both temperature and water availability are optimal) than on any other site type. Show all calculations and state your conclusion with regard to the original scientific question.

Name _____   Date _____

Answer Key for Chapter 11 Study Problems

1. **a.** Comparison-wise Type I error rate: **0.05**

   Experiment-wise Type I error rate: **0.2649**

   $$= 1 - (1 - \alpha)^{\text{Ntest}} = 1 - (0.95)^6$$

   If the allowable Type I error rate for each test of significance was set at $\alpha = 0.05$, the probability that the investigator would incorrectly conclude that this population was a different species, due only to random sampling variation, would be 0.2649.

   b.

Characteristic	$\bar{x}_1$	$\bar{x}_2$	Ntest	d	p-Value	Critical p
Anther length	1.8	1.2	6	0	0.001	0.00851
Petal length	3.4	3.0	6	1	0.01	0.01021
Sepal length	1.5	1.0	6	2	0.01	0.01274
Petal width	0.9	0.6	6	3	0.025	0.01695
Sepal width	0.5	0.4	6	4	0.15	
Pistil length	0.9	1.1	6	5	0.10	

   c. Anther length, Petal length, and Sepal length are all significantly different between these two populations of plants, with experiment-wise Type I error rate held to 0.05 or less.

   d. Therefore, the newly discovered population is a different species.

2. **a.** Treatment variable: **Site Type** (which is related to temperature and soil water)

   b. Response variable: **Mass Loss** (which is a measure of decomposition)

   c. The assumptions: *Randomized, unbiased study design:* Bags of uniform leaf litter mix were randomly assigned to different sites. Equivalent treatment groups. Assumption valid.

   Populations (sites) have equal variance in mass loss: $F_{\max} = 1.79^2 / 1.20^2 = 2.23$. Critical $F_{\max}$ (with $k = 4$ and df $= n - 1 = 10$) $= 5.67$. Since the computed $F_{\max}$ is less than the critical $F_{\max}$, we can assume the population variances are equal. Assumption valid.

   Population distributions are Normal: Minimal information is provided about shape of the data distribution. The medians and means are approximately equal, except for the upper-south group. With equal sample sizes for all groups, ANOVA would likely not be greatly affected by modest deviations from Normality.

   d. H_0: $\mu_{\text{LS}} = \mu_{\text{US}} = \mu_{\text{LN}} = \mu_{\text{UN}}$

   H_a: **Mean mass loss differs among at least two sites.**

   e. The probability that the observed differences among the group means is due solely to random variation is < 0.00001.

   f. *Conclusion:* The ANOVA provides strong evidence that mass loss differed among the four site types.

   g. $\text{HSD} = Q_{k=4, df=40} \sqrt{\text{MSE} / n_i}$

$$= 3.791 \sqrt{2.0/11}$$

$$= 1.62$$

Low-North	Upper-North	Upper-South	Low-South
4.48 a	5.07 a	5.38 a	7.92 b

Conclusion: Mean mass loss (decomposition) was greater on the low-south site than on the other sites. There were no other significant differences.

   h. $H_0: \mu_{LS} = \mu_{US} = \mu_{LN} = \mu_{UN}$

$H_a: \mu_{LS} > \mu_{US} = \mu_{LN} = \mu_{UN}$

$C = 1.0 \, \bar{x}_{LS} - 0.33 \, \bar{x}_{US} - 0.33 \, \bar{x}_{LN} - 0.33 \, \bar{x}_{UN}$

$$= 7.92 - 0.33 \, (5.38) - 0.33 \, (5.01) - 0.33 \, (4.48)$$

$$= 3.01$$

$S_c = \sqrt{\text{MSE} \, \{1/11 + 0.33^2/11 + 0.33^2/11 + 0.33^2/11\}}$

$$= \sqrt{2.0 \, \{0.1206\}}$$

$$= 0.49$$

$t_{test} = C / S_c = 3.01 / 0.49 = 6.14$

$p < 0.0005$

Conclusion: The results of this study provide strong evidence that decomposition rate is greater on warmer, moister lower south slopes than on any other site.

Exercise A: Bonferroni Procedure

Objective   You will be able to implement an analysis that requires multiple tests of significance and use the sequential Bonferroni procedure to control the overall experiment-wise Type I error rate. Specifically, you will:

1. Perform exploratory data analyses for multiple variables measured during a Before-After experimental design to determine if the data meet assumptions for the paired *t*-test.

2. Perform either the paired *t*-test or sign test, as appropriate for each variable.

3. Perform the Bonferroni procedure to determine which, if any, variables indicate a significant difference between Before and After measurements.

4. Draw appropriate conclusions with regard to the original scientific question.

Introduction   Some have proposed that ketogenic diets rich in protein and fat, but with very little carbohydrates (e.g., Atkins diet, the Zone diet, Protein Power diet), can be effective for weight loss, treating hyperinsulinemia (high blood insulin and high blood sugar), and modifying blood cholesterol composition to reduce risk of cardiovascular disease.

A researcher wanted to determine if reduction in blood sugar and LDL cholesterol attributed to these diets were simply consequences of weight loss associated with the diets, or actually a result of the diet. He implemented a Before-After study with $n = 12$ volunteer adult males at a university, including students, custodial staff, and professors. Subjects were measured for a variety of variables before beginning the ketogenic diet and again after six weeks on the diet. The diet was designed to virtually eliminate carbohydrates (< 30 g/day), but with added calories from fat to prevent weight loss. Response variables measured on the subjects included the following:

TC   Total blood cholesterol (Low values are considered "good.")

HDL   High-density lipoprotein cholesterol (sometimes called "good" cholesterol).

LDL   Low-density lipoprotein cholesterol (sometimes called "bad" cholesterol components).

TC/HDL   Ratio of total cholesterol to HDL cholesterol (This variable is used to assess risk of cardiovascular disease; low values are "good.")

%Fat   Percentage of total body weight that is accounted for by fat (Low values are considered "good.")

The following data table contains Before and After measurements for these variables on the 12 study subjects. The Before measurements have the suffix **0** (time zero) and the After measurements the suffix **6** (time at 6 weeks).

Analysis of these data requires a separate test of significance for each of the variables listed above. Given the Before-After study design, a paired *t*-test or sign test is the appropriate test of significance to determine if the data provide evidence that the ketogenic diet caused a change in the subjects' blood cholesterol or body fat. However, the multiple tests of significance all address the same overall Null hypothesis that the diet was not effective. Hence, concerns about cumulative Type I error rate should be addressed. When multiple tests of significance are required because there are multiple variables, the Bonferroni procedure is an appropriate means for controlling experiment-wise error rate.

The Data

TC0	TC6	HDL0	HDL6	LDL0	LDL6	TC/HDL0	TC/HDL6	%Fat0	%Fat6
165	181	40	54	116	118	4.13	3.39	25.3	19.7
173	197	36	42	123	140	4.87	4.68	25.1	19.4
225	228	43	43	166	175	5.30	5.39	31.0	29.1
168	149	50	52	100	89	3.40	2.89	22.9	16.1
133	110	35	31	66	63	3.84	3.61	20.5	16.0
123	197	53	91	59	100	2.32	2.17	9.7	8.8
159	166	66	69	86	89	2.43	2.40	22.9	18.7
174	186	47	51	103	124	3.70	3.65	12.3	12.3
202	189	48	50	136	122	4.24	3.82	20.1	15.7
111	144	42	57	60	81	2.67	2.53	12.4	12.9
163	145	52	49	99	89	3.16	2.99	20.9	18.0
186	182	58	47	104	125	2.81	3.12	22.9	15.9

Instructions

1. Enter these data into a computer statistics program. The data should be in the format shown above to allow for paired *t*-tests.

2. *Exploratory data analysis:*

   a. The assumptions for the paired *t*-test apply to data distributions for differences (*D* = After − Before). To assess if the assumptions are fulfilled, you must first compute these *D*-values between paired measurements *for each variable.*

   b. Perform the following exploratory analyses for the *D*-values of each variable:

      • Boxplot (good for identifying the presence of outliers).

      • Normal quantile plot (good for assessing the Normality assumption, especially when sample size is small).

      • Summary statistics, including sample size, mean, median, standard deviation, minimum, and maximum.

      Copy/Paste these analyses of the *D*-values for each response variable to a separate word processor page. *On each page, type your assessment as to whether or not the Normality assumption of the paired t-test is fulfilled.* Your assessment should take into account the data distribution, the sample size, and the Central Limit Theorem.

   c. Based on your assessment of the Normality assumption, perform either the paired *t*-test or the sign test on the Before and After measurements for each of the response variables. Copy/Paste each of these statistical print-outs to the bottom of the word processor page below the exploratory data analyses for each variable.

   d. Set up a table that displays results of the tests of significance in a format appropriate for the sequential Bonferroni procedure (see Chapter 11 in the textbook for examples). Calculate the adjusted critical *p*-value for each row (test of significance) in the table.

Report   Your report should include statistical analyses for each variable on a separate page (five pages of print-out in all), as well as a table that summarizes the results for all tests of significance and includes the Bonferroni critical *p*-value for each test. *Type answers* to the following questions.

1. Suggest reasons why the investigator who did this study used a Before-After study design instead of a completely randomized study design with control and treatment groups.

2. **a.** What parts of the results from all of these analyses represent Comparison-wise error rate?

   b. What is the problem with using the values for Comparison-wise error rate as the basis for drawing conclusions about whether or not the ketogenic diet affected blood cholesterol or percentage of body fat?

   c. What is the Experiment-wise error rate for your sequential Bonferroni analysis?

3. State your conclusions with regard to whether or not the ketogenic diet affected blood cholesterol variables or percentage of body fat. These conclusion statements should be in a format suitable for a scientific paper. That is, where differences are significant, state the nature of the difference (increase or decrease), and include the appropriate *p*-value.

4. Given the study description, describe the larger population to which you think the results of this study apply.

Exercise B: Analysis of Variance

Objectives   1. You will be able to use descriptive summary statistics, boxplots, and Normal quantile plots to verify that data meet the assumptions for Analysis of Variance.

2. You will perform ANOVA using computer statistics software, including entering data in the appropriate format.

3. You will perform Tukey's HSD multiple comparisons test.

4. You will perform the Kruskal-Wallis test and a multiple comparisons test for medians.

Introduction   For this exercise, you will be analyzing total stem wood volume for trees in four plantations that differed with regard to spacing between trees and tree species: (1) Red pine planted at 4ft spacing (**RP4**), (2) Red pine planted at 6ft spacing (**RP6**), (3) Red pine planted at 8ft spacing (**RP8**), and (4) Scotch pine planted at 4ft spacing (**SP4**). The three red pine plantations were part of a study to determine the optimum spacing to maximize individual tree growth for the greatest number of trees per acre. Trees planted at wider spacing suffer less from competition for light, water, and nutrients, and hence should grow faster and become larger than trees planted at close spacing. However, this increase in growth with wider spacing occurs only within the range of spacings that individual trees are capable of filling with their roots and leaves. At wider spacing, trees would not grow any faster, but there would be fewer trees per acre and so less total wood produced. The Scotch pine plantation was included to determine if this species has greater growth than red pine on sandy and nutrient-deficient soils. In general, Scotch pine is known to be better adapted to infertile soils than red pine. In this part of the study, the investigator wanted to determine if Scotch pine grows better than red pine regardless of spacing. The data are total wood volume (dm^3) in the stems of individual trees randomly sampled from the adjacent experimental plots. These data were obtained when the trees were 50 years old. Greater total wood volume at age 50 would indicate higher growth rates during the previous 50 years.

In this exercise, you will enter the data, perform graphical and statistical analyses to determine if the data meet the assumptions for Analysis of Variance, use the log-transformation to adjust the data for unequal variances and non-Normal distributions, and then perform ANOVA. You will also perform the Kruskal-Wallis test for equality of medians. You will perform multiple comparison test to determine which group means or medians are significantly different at the $p \leq 0.05$ level.

Entering Data for ANOVA   Enter the data as follows. Create two columns of data, one titled **Group** and the other **Volume**. Enter all the tree stem wood volume values in Table 11.1 below *in a single column* (stack the data for all groups/species). For each data value in the variable **Volume**, enter a corresponding value for **Group** on the same line in the **Group** variable (values for this categorical variable should be **RP4**, **RP6**, **RP8**, and **SP4**). After you enter all data values, select **File/Save File As...** and save your data file.

TABLE 11.1   Data for stem wood volume for red pine and Scotch pine.

Group				Data Format for ANOVA	
RP4	RP6	RP8	SP4	Group	Volume
88	131	234	287	RP4	88
95	172	257	223	↕	↕
104	116	183	333	RP4	108
131	205	118	273	RP6	131
122	229	200	193	↕	↕
143	151	124	253	RP6	163
120	199	104	188	RP8	234
86	180	135	148	↕	↕
114	168	133	174	RP8	177
126	151	148	172	SP	287
86	171	151	202	↕	↕
107	216	200	195	SP	202
156	169	141	333		
96	146	232	173		
108	163	177	202		

Determine If Data Meet the Assumptions for Analysis of Variance

Assumptions for analysis of variance are:

1. The population distributions for the variable are Normal.

2. The data for all groups come from populations that have the same variance.

3. The data were obtained by randomized, unbiased study design.

We know assumption (3) is fulfilled based on the description of the study in the Introduction. You must determine if the Volume data meet the other assumptions.

1. To assess whether or not the data distributions for the four groups are approximately Normal, produce side-by-side boxplots (to identify outliers easily) and separate Normal quantile plots for the data values in each group.

2. Generate a separate set of summary statistics for each group (including the group name, mean, median, standard deviation, and sample size). Comparisons of means vs. medians for each group can help you assess if outliers have undue influence on the group mean. The standard deviations are needed to assess the equal variances assumption. The mean and standard deviation values are used to compute confidence intervals for group means. Finally, these summary statistics will be useful if you need to perform a multiple comparisons test.

3. If your statistics software does not provide a test for equality of variances as part of the ANOVA procedure, you will need to perform the F_{Max} test by hand calculation to determine if data meet the equal variances assumption.

4. When the data fail to meet the Normality and/or equal variances assumption, one option for analysis is to transform the data so that the assumptions are fulfilled and then apply ANOVA to the transformed data values. Create a new variable that contains the log-transformed data values. Give the variable the name **LogVol**. Generate side-by-side boxplots, Normal quantile plots, and summary statistics by Group for the LogVol transformed data.

5. When the Normality assumption is violated, another option for analyzing data such as these is to perform the nonparametric Kruskal-Wallis test and associated multiple comparisons test for medians. If your statistics software does not provide the Kruskal-Wallis test in its menu, you will need to create a new variable that contains the ranks of the Volume data values. Give this variable the name **RankVol**. *Note:* Most statistics and spreadsheet software provide a function that will automatically determine the ranks for data values, often in the form **Rank**(*variable name*). If you must use the ANOVA version of the Kruskal-Wallis test, you should generate summary statistics by Group for the RankVol data to obtain the mean rank values for each group. You will need mean ranks for the multiple comparisons test for medians. If your statistics software offers the Kruskal-Wallis test as a menu choice, the print-out from this test will usually include values for the mean ranks by group.

Performing Analysis of Variance

1. Perform Analysis of Variance for the original Volume data. This will usually entail specifying **Group** as the treatment (X) variable and **Volume** as the response (Y) variable. For this completely randomized experimental design, you should choose **One-way ANOVA** if you are offered a choice of different kinds of ANOVA.

2. Perform ANOVA for the log-transformed data in variable **LogVol**. Again, specify Group as the treatment (X) variable.

3. If your statistics software offers the Kruskal-Wallis test in its statistics menu, perform this test on the original **Volume** data. If the Kruskal-Wallis test is not offered as a menu item, perform ANOVA on the variable **RankVol**. *Note:* Performing ANOVA on the ranks of the data values is equivalent to performing a Kruskal-Wallis test. Even though the print-out from this "ANOVA on rank values" looks like the other ANOVA analyses, it is the nonparametric Kruskal-Wallis test.

Report   The print-outs from statistical analyses should be arranged on pages as described below:

Page 1: Include summary statistics, side-by-side boxplots, and the four Normal quantile plots (one for each group) for the original Volume data. Put the following title on the top of the page: **Exploratory Data Analysis for Volume**, and put your name in the upper-left corner. Show the calculations for the F_{max} test of equality of variances on this print-out, just below the summary statistics, and type your assessment regarding the equal variances assumption.

Page 2: Include the ANOVA print-out for the analysis of the original Volume data. Below this print-out show the calculations for the Tukey HSD test. List the group means in order (with the Group ID label next to each) and use either the letter or underline method to indicate which means are significantly different at the $p < 0.05$ level. Type your conclusions with regard to the original scientific questions described in the Introduction.

Page 3: Include summary statistics, side-by-side boxplots, and the four Normal quantile plots (one for each group) for the transformed data in **LogVol**. Put the following title on the top of the page: **Exploratory Data Analysis for Log(Volume)**, and put your name in the upper-left corner. Show the calculations for the F_{max} test of equality of variances on this print-out, just below the summary statistics, and type your assessment regarding the equal variances assumption.

Page 4: Include the ANOVA print-out for the analysis of the transformed data in **LogVol**. Below this print-out show the calculations for the Tukey HSD test. List the group means in order (with the Group ID label above each) and use either the letter or underline method to indicate which means are significantly different at the $p < 0.05$ level. Type your conclusions with regard to the original scientific questions described in the Introduction.

Page 5: Include the Kruskal-Wallis test print-out *or* summary statistics by Group of RankVol and the ANOVA print-outs for the analysis of RankVol. Below this print-out, show the calculations for the multiple comparisons test for medians. List the group medians (from summary statistics print-out reports) in order (with the Group ID label above each median) and use either the letter or underline method to indicate which medians are significantly different at the $p < 0.05$ level.

Answer the following questions in *complete sentences and paragraphs* that are under-standable without referring back to the questions. *Your answers should be typed.*

1. Do the original data meet the assumptions for ANOVA? For each assumption, state your interpretation of the data analysis, with appropriate specific references to summary statistics and graphics. State your conclusion with regard to whether or not it is valid to analyze these data using ANOVA.

2. Do the log-transformed data meet the assumptions for ANOVA? Again, state your interpretations of the data analysis with specific reference to those print-outs that provided the basis for your statements.

3. **a.** Interpret the meaning of the *p*-value from the ANOVA of the original Volume data in the context of this study.

   b. Explain the reasoning behind your use of this *p*-value in arriving at your conclusion.

4. Compare the results from the three tests (ANOVA on the original Volume data, ANOVA on log-transformed data, Kruskal-Wallis test).

   a. What does this comparison indicate to you about the robustness of ANOVA to violation of assumptions when sample sizes are equal?

   b. Compare the results from the three multiple comparisons tests. Explain any differences.

5. Based on the *most appropriate* of the three analyses you performed, interpret the results with regard to the original biological questions described in the first paragraph of the Introduction.

   a. Does increasing spacing result in increased growth of red pine? What is the optimal spacing to get maximum growth on the largest number of trees?

   b. Does Scotch pine grow better than red pine on this sandy, infertile soil?

6. Perform a planned contrast on the LogVol data to test the hypothesis that Scotch pine grows better than the red pine groups on this infertile site. *Remember:* The pooled estimate of the variance you will need for these computations is equal to the Mean Square Error term in the ANOVA table. Show all computations, the resulting *t*- and *p*-values, and state your conclusion.

c. Compute the overall proportions of individuals in each ASEPSIS class for all three treatment groups combined and enter these values in the spaces provided in the table above.

d. Compute the expected values and enter them in the cells of the table above.

e. Compute the $(O - E)^2/E$ values and enter them in the cells of the table above.

f. Compute the χ^2_{test} statistic and determine the df and the p-value.

χ^2_{test} _____   df _____ p-value _____

g. State your conclusion with regard to the overall Null hypothesis.

h. *For the ASEPSIS class that has the largest sum of* $(O - E)^2/E$ *values across the three treatment groups*, perform a Tukey-style unplanned multiple comparisons test to determine which of the treatment groups differ with regards to the proportion of patients in that ASEPSIS class. *Note:* If you were doing this analysis using a computer, you would use the subset χ^2 analysis to determine which ASEPSIS classes had significant differences among the three treatment groups. Since you are doing this calculation by hand, we will simply do the multiple comparisons test for the one ASEPSIS class most likely to have different proportions of individuals among the three groups.

i. State your conclusion based on the results of the multiple comparisons test. What do these results mean with regard to using warming to reduce the incidence of post-surgical infection?

Chapter 12 Supplemental Problem

1. Concerns have been raised that the proportion of women who give birth to their babies by surgical C-section has been increasing. Some propose that this may be due to the increased use of narcotics and epidural to relieve the pain associated with childbirth. They hypothesize that pain-relieving drugs may slow labor or inhibit the mother's ability to push the baby through the birth canal. This could make it more likely the mother will require a C-section than women who practice natural childbirth methods. A study of hospital records for all births in a hospital during the last year compared the proportion of women who required a C-section between three groups of women: (1) those who received an epidural, (2) those who received a narcotic pain reliever, and (3) those who practiced natural childbirth methods. The results of this study are presented in the following table.

Experimental Groups				
Outcome	Epidural	Narcotic	Natural Methods	Total
C-section	97	67	44	208
Exp	_____	_____	_____	$\hat{p}$ = _____
$(O - E)^2/E$	_____	_____	_____	
Vaginal Birth	1086	1119	810	3015
Exp	_____	_____	_____	$\hat{p}$ = _____
$(O - E)^2/E$	_____	_____	_____	
Total	1183	1186	854	3223

   a. Compute the 95% confidence intervals for the proportion of women who practiced natural birth methods but then had a C-section (P_{Nat}).

b. Explain to someone who knows no statistics why we can't just say that 5.15% of women who attempt to deliver their baby by natural birth have a C-section?

c. Explain why it would be inappropriate to use multiple two-sample Z-tests to determine which of the three groups differ with regard to the proportion of women who delivered by C-section.

d. Perform a χ^2 test of homogeneity to determine if the proportion of women who had C-sections differs among any of the three groups of women:

(1) Hypotheses:

H_0: _____

H_a: _____

(2) Assess whether or not these data fulfill the assumptions for this test of significance.

(3) Compute the expected values for each cell in the table and write these in the first space provided in each cell of the table.

(4) Compute the quantities $(E - O)^2/E$ for each cell in the table and write these values in the second space provided in each cell of the table.

(5) Compute the X^2 statistic, and determine the degrees of freedom and the approximate p-value using Appendix Table 8.

χ^2 _____   df _____ p-value _____

Use a computer spreadsheet program to obtain the exact p-value. _____

(6) State your conclusion with regard to the original scientific question.

(7) Perform a Tukey-style unplanned multiple comparisons test to determine which of the three groups of women differ with regard to the proportion of women who delivered by C-section.

Chapter 12 Study Problem

1. A pregnant woman wants very much to deliver her baby by natural childbirth. She is concerned that in these days of soaring medical malpractice claims some obstetricians and hospitals avoid this risk by performing C-section delivery at the first hint of trouble in the birth process. She wants to avoid delivering her child at a hospital with rules or practices that might increase the chance she would have to deliver her baby by C-section. She lives in a large city that has four hospitals (A–D). She makes inquiries about the number of births at each hospital in the past year, and how many were done by C-section. The table below displays the results. Do these data provide any evidence that some of these hospitals have a higher or lower proportion of births by C-section than the others? Implement a χ^2 test of homogeneity.

	Hospital				
Outcome	A	B	C	D	Total
Vaginal Birth	740	3100	1130	1930	6900
Exp	_____	_____	_____	_____	
$(O - E)^2/E$	_____	_____	_____	_____	_____
C-Section	20	300	120	80	520
Exp	_____	_____	_____	_____	
$(O - E)^2/E$	_____	_____	_____	_____	_____
Total	760	3400	1250	2010	7420

a. Evaluate whether or not these data meet the assumptions for this test.

b. State the Null and Alternative hypotheses.

   H_0: _____

   H_a: _____

c. Compute the expected counts for each cell and enter these on the first blank line in each cell of the table.

d. Compute the values $(O - E)^2/E$ values for each cell and enter these on the second line in each cell of the table.

e. Compute the χ^2_{test} statistic, determine the degrees of freedom, and determine the approximate p-value using Appendix Table 8.

χ^2_{test} = _____   df = _____ $p <$ _____

f. Based on this p-value, state your conclusion with regard to the woman's original question.

g. Perform a Tukey-style unplanned multiple comparisons test to determine which hospitals differ with regard to the proportion of births done by C-section.

h. Based on the multiple comparisons test, which hospitals have lower proportions of births by C-section?

Answers to Chapter 12 Study Problems

1. a. Assumptions:

The data were obtained for all births at these hospitals in the last year. Presumably, the data for the last year are representative of birth outcomes at these hospitals in recent years.

The population of interest here is "All women who give birth," and is extremely large.

Average expected frequency is $N / (\text{\#rows})(\text{\#cols}) = 742 / (2)(4) = 92.75 \geq 6$. The assumption about the minimum sample size is valid.

b. $H_0: P_{N,A} = P_{N,B} = P_{N,C} = P_{N,D}$

$ P_{C,A} = P_{C,B} = P_{C,C} = P_{C,D}$

H_a: The proportion of vaginal and Caesarian births differs between two or more hospitals.

c. and **d.**

Outcome	A	B	C	D	Total
Vaginal Birth	740	3100	1130	1930	6900
Exp	**706.8** (760×0.93)	**3162** (3400×0.93)	**1162.5** (1250×0.93)	**1869.3** (2010×0.93)	$\hat{p} = \mathbf{0.93}$
$(O - E)^2/E$	1.6	1.2	0.9	2.0	
C-Section	20	300	120	80	520
Exp	**53.2** (760×0.07)	**238** (3400×0.07)	$E = \mathbf{87.5}$ (1250×0.07)	**140.7** (2010×0.07)	$\hat{p} = \mathbf{0.07}$
$(O - E)^2/E$	20.7	16.2	12.1	26.2	
Total	760	3400	1250	2010	7420

Table title: **Hospital**

e. $\chi^2 = 1.6 + 1.2 + 0.9 + 2.0 + 20.7 + 16.2 + 12.1 + 26.2 = \mathbf{80.9}$

$\text{df} = (\text{\#rows} - 1)(\text{\#cols} - 1) = (1)(3) = 3$

$P[\chi^2 \geq 80.9 \text{ with df} = 3 \text{ if } H_0 \text{ is true}] < \mathbf{0.0005}$

f. *Conclusion:* The proportion of births by Caesarian section differs among the four hospitals ($p < 0.0005$).

g. Multiple comparisons test ($\hat{p}$ for C-section births)

	A	B	C	D
$\hat{p}$	0.026	0.088	0.096	0.04
$\hat{p}'$	9.28	17.26	18.05	11.54

Comparison	Difference	n_1	n_2	SE	$Q_{0.05,\,4}$	MSD	Conclusion
A − B	−7.98	760	3400	0.813	3.633	2.95	Sig. Diff.
A − C	−8.77	760	1250	0.932	3.633	3.39	Sig. Diff.
A − D	−2.26	760	2010	0.863	3.633	3.14	Not Sig. Diff.
B − C	−0.79	3400	1250	0.670	3.633	2.43	Not Sig. Diff.
B − D	5.72	3400	2010	0.570	3.633	2.07	Sig. Diff.
C − D	6.51	1250	2010	0.729	3.633	2.65	Sig. Diff.

h. *Conclusion:* A smaller proportion of births at hospitals A and D were by C-section than at hospitals B and C. The difference in the proportion of C-section births between hospitals A and D was not significant, nor was the difference between hospitals B and C.

Exercise A: Two-Way Table Analysis

Objective

1. You will perform two-way table analysis to test hypotheses regarding the homogeneity of proportions of individuals that fall into classes of a single categorical variable among two or more populations.

2. When the overall Null hypothesis of equal proportions is rejected, you will perform subset χ^2 analysis to identify the nature of the differences.

Introduction

Severe droughts often injure trees in ways that reduce their capacity for growth over some period of years after the actual drought is over. During this period of reduced growth after droughts, oak trees may be more susceptible to attack by a number of insect pests that are capable of killing only weakened trees. This phenomenon is similar to the opportunistic attacks on AIDS patients and the elderly by a variety of disease organisms that do not usually affect healthy people. The "response" variable in this research was **Decline Duration**, defined as the number of consecutive years during and after a severe drought in 1953–54 when tree growth was below the average growth for the 5-year period prior to the drought. The larger the value of this variable, the longer the tree suffered reduced growth and perhaps increased vulnerability to disease and death. Hence, individuals with large values for this variable were more negatively affected by the drought than individuals with small values (which indicate rapid recovery). The specific research question addressed by this research is whether or not there is a difference between white oak (*Quercus alba*) and black oak (*Quercus velutina*) in their growth response to this severe drought. Previous research has reported that more black oaks than white oaks die after severe droughts. The data you will analyze for this exercise were collected to determine if this difference in mortality rates might be explained by different responses to drought.

Study Design

A sample of $n = 120$ white oak and $n = 110$ black oak were cored to obtain tree-ring width data. The sample trees were located across a range of local site conditions, as characterized by topographic position and soil pH. For this exercise, we will focus solely on a comparison between the two oak species.

Instructions

1. The data file includes the Chapter 12 Exercise variables, **Species** code (WO = white oak, BO = black oak) and post-drought **Decline** duration (yrs).

2. The variable **Decline** contains the exact number of consecutive years during and after the drought that the individual trees exhibited below average growth. Each distinct value of this variable would constitute a "class." If you attempted to perform two-way table analysis using the original data, the resulting table would have 20 to 30 rows, with few or no trees in most rows. To analyze these data with a two-way table approach, we must first convert a more or less continuous, quantitative variable into a numeric categorical variable with a relatively small number of discrete classes. This is done either at the time the data are entered into the data file *or* it can be done by converting the continuous data values into a categorical format. To do this latter procedure with

the Decline data, use the following formula to compute a new variable "Decline Class" (shortened to **Dclass**):

Dclass = < bin width > × **Truncate** (**Decline** / < bin width >) + < bin width > /2

In this equation, "**bin width**" refers to the width of the classes. For example, we might convert the Decline Duration variable into a quantitative categorical variable with a bin width of five years for each class (e.g., 0 to 4.9, 5 to 9.9, 10 to 14.9, etc.). **Truncate** is a generic reference to a function that cuts-off decimal places and rounds down to the nearest whole number. This function has various names in different spreadsheet and statistics software (e.g., INT, TRUNC, RoundDown). Below, I provide some examples of how this formula might be implemented for different bin widths and with different data values.

Bin Width	Data Value	Formula	Decline Class Mid-Point
5	4	5*TRUNC(4 / 5) + 5/2 = 5 (0) + 2.5	2.5
5	21	5*RoundDown(21 / 5) + 5/2 = 5 (4) + 2.5	22.5
10	13	10*TRUNC(13 / 10) + 10/2 = 10 (1) + 5	15
10	7	10*RoundDown(7 / 10) + 10/2 = 10 (0) + 5	5

Create this new quantitative categorical variable using a bin width of 10 years and give it the name **DClass**.

3. Before performing two-way table analysis, you should select the kinds of information you want in your table. You should always include the observed count, expected count, and standardized residual. In addition, for a χ^2 test of homogeneity of proportions (among different populations), you will also want the percentage (or proportion) of individuals in each decline class by species. If you stipulate the variable Species as the column variable of the table, you should request the "percent of column" in each cell of the table. You should also specify you want the row margins, column margins, and the χ^2 test statistics.

4. After specifying the format, implement the two-way table analysis with Species as the column variable and Dclass as the row variable.

5. If the overall χ^2 test is significant ($p \leq 0.05$), perform a subset χ^2 analysis to determine which Decline class categories differ between the two species. This is done by progressively deleting all data for specific decline classes with the largest standardized residuals, then rerunning the χ^2 analysis to determine if a significant difference still exists among the remaining decline classes.

   a. Deleting data for an entire decline class is most easily done if the data are sorted by decline class. Use the Sort function of your spreadsheet or statistics software, with the sorting controlled by the data values in the variable **Dclass** and all other variables in the data set included in the sort.

b. Look at the original two-way table for the species vs. decline class analysis (see Instruction 4 above) and determine which decline class has the largest sum of standardized residuals across the two species. Then locate the block of rows in the data for this Decline class and delete those rows. Make sure you have saved a back-up of your original data file before you begin to delete data.

c. Rerun the two-way table analysis of species vs. Dclass. If the p-value for the χ^2 test is no longer significant, then you can conclude that the only difference between the two species was with regard to the one deleted decline class.

d. If the p-value for the χ^2 test is still significant, repeat steps (b) and (c); select the decline class with the largest standardized residuals in the subset two-way table produced in step (c).

When the p-value for the χ^2 test goes from being significant to being not significant, you can conclude that the differences between the two species were with regard to the decline classes that were deleted.

e. Record the results of this analysis in a table with the following information:

Decline Classes in the Analysis	χ^2	p-Value
All	____	____
Class ____ Dropped	____	____
Classes ____ and ____ Dropped	____	____

Report Print the original two-way table for **Species** vs. **Dclass**, with your name and an appropriate title for the analysis at the top of the page. *Type answers* to the following questions.

1. Write the Null and Alternative hypotheses. Use appropriate subscripts to specify populations and variables.

2. Based on the overall χ^2 test with all classes of Dclass included, state your conclusion with regard to the original scientific question.

3. a. Based on subset χ^2 analysis, for what decline duration classes are there significant differences between white oak and black oak?

   b. Which species has a greater/lower proportion of individuals in those classes?

   c. Are these differences consistent with reports in the literature that more black oak die after droughts than do white oak?

2. Lead poisoning is a great public health problem for children living in older urban areas. Prior to the 1970s, most house paint contained lead. When the health risks of environmental lead were identified, its use in paint was banned. However, many older houses still have lead-based interior paint, which is now deteriorating to produce flakes and dust. Small children consume the flakes because they have a sweet taste. Paint dust contaminates surfaces; children touch them and then transfer the dust to their mouth. Public health researchers performed a study to determine if the abundance of older houses in urban neighborhoods was associated with a higher incidence of lead poisoning in young children who live there. The city was subdivided into census tracts, and the percentage of houses in each tract that were built prior to 1970 was determined. Researchers obtained records from a lead-poisoning screening of 27,590 children, and determined the percentage of children who live in each census tract with blood lead levels of 20 μg/dL or greater (definition for lead poisoning).

The Data

%Houses	0	5	10	21	27	36	41	50	53	57	60	63	70	78	98
%Children	.05	.06	.03	.4	.35	1.5	2.2	2.7	4.0	5.1	6.2	7.3	5.2	6.1	5.9

a. Enter these data into your statistics program.

b. Perform graphical analyses to determine if the data meet all assumptions for using Pearson's correlation coefficient r to measure the strength of the association (see the Chapter 13 Tutorial). Copy/Paste these graphics into a word processor and *type your assessment as to whether or not Pearson's r is a valid measure of the association.*

c. Use the statistics program to compute Pearson's r for the association between these two variables. *Note:* Use the program even though this may be invalid because the data violate one or more assumptions. The purpose is to demonstrate the consequences of using an invalid correlation coefficient.

d. Use the statistics program to compute Spearman's r_s. If your software does not offer this option, create two new variables that contain the ranks for X and Y and apply the Pearson's correlation analysis to the ranks to obtain Spearman's r_s.

e. Copy/Paste the two correlation print-outs to the same word processor page as the graphics you produced to assess whether or not the data met the assumptions for Pearson's correlation coefficient.

f. *Type answers* to the following questions on that page.

(1) Which of these two variables is the "cause" (X) and which is the "effect" (Y)?

(2) Describe the nature of the association between the percentage of houses built before 1970 and the percentage of children with lead poisoning.

(3) Is it valid to use Pearson's r to measure the strength of association between these two variables? Explain your answer.

(4) Is it valid to use Spearman's r_s to measure the strength of association? Explain.

(5) Compare Pearson's r and Spearman's r_s. What is the consequence of using Pearson's r when one or more of the assumptions are violated?

3. The data file **%BodyFat** contains data that were obtained from $n = 252$ volunteer men who were over 25 years old. The objective of the study that generated these data was to identify easily measured body dimension variables that might have a strong association with the percentage of body fat (i.e., the percentage of body mass that is fat). Perform correlation analysis to determine which body dimension variable in this data set has the strongest association with the percentage of body fat.

    a. Select **PctFat** and all body dimension variables (do not include Age or Density).

    b. Request that the software compute Pearson's r. This will produce a matrix of correlation coefficient values for all pairs of variables.

    c. If available, request that the software compute Spearman's r_s to obtain a similar matrix of all pair-wise rank correlations.

    d. These two correlation matrices may be too large to paste into a word processor page. Scan through the matrices on the screen to identify the one body dimension variable in each matrix that has the strongest correlation with percentage of body fat.

    e. Perform the exploratory data analyses required to assess if the data for PctFat and the body dimension variable most strongly correlated with PctFat meet the assumptions for Pearson's r. Paste these graphics onto a single word processor page and append that page to this one.

    f. Type answers to the following questions below the print-outs. Your answers should be in complete sentences.

        (1) Which body dimension variable had the strongest association with the percentage of body fat based on Pearson's r? Based on Spearman's r?

        (2) Do the data for PctFat and this variable meet the assumptions for Pearson's r? Explain with specific references to the graphical analyses you performed.

        (3) Compare Pearson's r and Spearman's r_S. Do these correlations give different or similar results? Suggest an explanation for this outcome.

4. Recent studies describe a trend of decreasing sperm counts in human males over the period 1938 to 1990 (Figure 13.1). These studies suggest decreased sperm count is coincident in time with increased use of estrogen-like pesticides. These pesticides have been shown to bind to the human estrogen receptor (hER) protein in yeast cells that were bio-engineered to produce human hER. During human development, when estrogen binds to the hER receptor, this "turns-on" a number of genes. If these genes are activated at inappropriate times during development, or in inappropriate amounts, the results can include increased risk of breast cancer and malformation of male sex organs. Toxicological studies performed at the time these estrogen-like pesticides were certified indicated minimal effect of each pesticide on test organisms. However, results from a recent study (Arnold et al. 1996) demonstrated that exposure to combinations of these pesticides can act synergistically to increase their binding to the human estrogen receptor hER protein.[1]

---

[1]Arnold, S. F., Klotz, D. M., Collins, B. M., Vonier, P. M., Guillette, L. J., and McLachlan, J. A. 1996. Synergistic activation of estrogen receptor with combinations of environmental chemicals. *Science* 272:1489–1492.

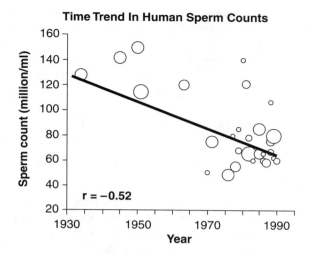

Time Trend In Human Sperm Counts

r = −0.52

FIGURE 13.1   **Results of 30 studies of human sperm counts performed over the period 1938 to 1990.** The studies were done across a number of countries. Estrogen-like pesticides were introduced after 1960. Each circle in the graph represents the average sperm count for one study. The size of the circles represents the sample size (number of men) measured in the study. This figure is based on data presented by Sharpe and Skakkenbaek (1993) *Lancet 341*:1392–1395.

a. Based on this information, evaluate how well current knowledge meets the three criteria for concluding that there is a cause-effect relationship between decreasing sperm counts in human males and estrogen-like pesticides.

b. Study the graph above. Describe the nature and strength of the association between sperm counts and time (years).

c. How does the size of the circles influence your assessment about using these data to determine if there is an association (trend in sperm counts over time)?

Name _____   Date _____

5. A student in an animal physiology class did an experiment to determine the effect of environmental temperature on the heart rate of leopard frogs. She obtained 11 frogs of approximately the same age, size, and gender, and randomly assigned each animal to a container kept at a temperature between 3°C and 33°C. After the frogs had equilibrated to their ambient temperature, she measured their basal heart rate (BHR). The table below presents her results.

Frog ID#	1	2	3	4	5	6	7	8	9	10	11
Temperature	3	6	9	12	15	18	21	24	27	30	33
BHR (beats/min)	3	6	14	13	20	23	28	27	33	38	42

The following is a print-out that displays the results of a linear regression analysis of the association between temperature (Temp) and basal heart rate (BHR). *Note:* The Y-intercept term is called the "Constant" in the regression equation, and "Coefficient" refers to the slope and Y-intercept values of the regression equation.

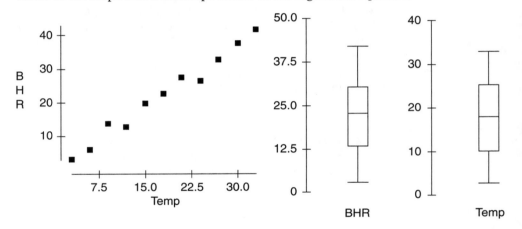

Summary Statistics

Variable	Count	Mean	Median	Variance	StdDev	Min	Max
BHR	11	22.4545	23	160.273	12.6599	3	42
Temp	11	18	18	99	9.94987	3	33

Dependent variable is: BHR
R-squared = 98.2%   R-squared (adjusted) = 98.0%
s = 1.81 with 11 − 2 = 9 degrees of freedom

Source	Sum of Squares	df	Mean Square	F-ratio
Regression	1573.24	1	1573.24	480
Residual	29.4909	9	3.27677	

Variable	Coefficient	s.e. of Coeff	t-ratio	Prob
Constant	−0.236364	1.171	−0.202	0.8445
Temp	1.26061	0.05753	21.9	≤ 0.0001

y-intercept (Constant)
slope (Temp)

Describe what the slope coefficient value represents *in this context*.

f. Explain the interpretation of the *p*-value associated with the *Y*-intercept (constant), and state your conclusion with regard to the *Y*-intercept for the LSR equation relating percentage of body fat to abdominal circumference.

g. Explain the interpretation of the *p*-value associated with the slope coefficient, and state your conclusion with regard to the slope for the LSR equation relating percentage of body fat to abdominal circumference.

h. Based on the results of this regression analysis, state your conclusion with regard to the scientific question as to the nature of the association between percentage of body fat and abdominal circumference for men. Use a format appropriate for a scientific paper and include the appropriate *p*-value from the regression print-out.

i. The Analysis of Variance table for this regression analysis presents an *F*-value and associated *p*-value. What Null and Alternative hypotheses are being tested by this *F*-test in the context of this specific research project?

j. What do you conclude from the ANOVA table results? State your conclusion in the context of this specific scientific question. Use a format suitable for a scientific paper.

k. Report the value of the coefficient of determination (Adjusted R-squared) and interpret this statistic in the context of the original scientific question. Use terms that are understandable to the general public.

7. a. Use the LSR equation you developed in Problem 6, PctFat $= -39.3 + 0.63$ (Abdomen), to compute predicted mean percentage of body fat for individuals whose abdominal circumference is 50 cm.

b. Use this equation to compute the predicted percentage of body fat for individuals whose abdominal circumference is 250 cm.

c. Do you see any problems with these predicted percentage of body fat values? If so, explain the cause of these problems.

d. How would you avoid the problem you described above?

8. Using information from the regression print-out for the analysis of the association between the percentage of body fat and abdominal circumference, compute the correlation coefficient for this association. Show your work.

9. **a.** Based on the regression analysis for the association between % BodyFat and abdominal circumference, compute the 95% confidence interval for ($\mu_{Y|X}$) = mean percentage body fat for the abdominal circumference value **90 cm** and for the value **120 cm**. Show all work. Report each of these two intervals as $\hat{y} \pm$ margin of error.

b. Explain the interpretation of these confidence intervals. Use terms understandable to the general public.

c. Compare the margin of error values of the confidence intervals for $(\mu_{Y|X})$ for the two abdominal circumference values **90** and **120**. Explain any differences. (*Hint:* You will need to refer to the exploratory data analyses.)

d. Compute the 95% *prediction* interval for $(Y|X)$ = an individual person's percentage of body fat for abdominal circumference values **90** and **120** cm. Show all work. Report these intervals as $\hat{y} \pm$ margin of error.

e. Explain the interpretation of these prediction intervals. Use terms understandable to the general public.

f. Compare the margin of error values between the confidence intervals computed in part (a) and the prediction intervals for the same abdominal circumference values computed in part (d). Explain any differences.

10. In Problem 5, R^2 (adjusted) for the regression of leopard frog heart rate on environmental temperature was 98% (0.98). Explain the meaning of this statistic in this context. Use terms understandable to the general public.

11. a. Based on the regression analyses for the association between temperature and frog basal heart rate presented in Problem 5, compute the Pearson's correlation coefficient for this association and perform a test of significance for this correlation. (Describe the sampling distribution, compute the t-test statistic, and determine the p-value.)

b. Compare the t-value and p-value for the correlation coefficient to the t-value and p-value for the slope coefficient presented in the table of coefficients in the regression print-out for Problem 5. What do you conclude about correlation and regression analyses?

Name _____   Date _____

Chapter 13 Supplemental Problem

1. Suppose that you wanted to estimate a man's percentage of body fat based on his weight. The following is the print-out of a regression analysis of the association between these two variables.

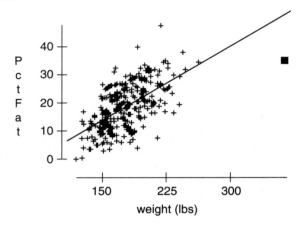

Summary Statistics

Variable	Count	Mean	StdDev	Min	Max
PctFat	252	19.1508	8.36874	0	47.5
Wt	252	178.924	29.3892	118.5	363.15

Dependent variable is: PctFat
R-squared = 37.5% R-squared (adjusted) = 37.3%
s = 6.629 with 252 – 2 = 250 degrees of freedom

Source	Sum of Squares	df	Mean Square	F-ratio
Regression	6593.02	1	6593.02	150
Residual	10986	250	43.9439	

Variable	Coefficient	s.e. of Coeff	t-ratio	Prob
Constant	–12.0516	2.581	–4.67	≤ 0.0001
Wt	0.174389	0.01424	12.2	≤ 0.0001

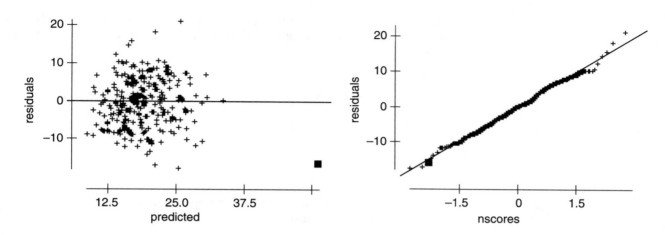

a. Is there an association between weight and percentage of body fat? Explain how you arrived at your conclusion. Make specific reference to parts of the print-out.

b. Based on the summary statistics, describe the men that made up this sample.

c. Suppose that one of the men included in this sample weighed 150 pounds and had seven percent body fat. Compute the predicted value for his percentage of body fat and the residual. How might you describe this man? (This is a biological question.)

d. What do you conclude from the t-test for the Y-intercept in the table of coefficients at the bottom of the regression print-out? How do you interpret this result in biological terms?

e. What do you conclude from the t-test for the slope coefficient? How do you interpret this result in biological terms?

f. What can you conclude from the ANOVA F-test?

g. Explain the meaning of $R^2 = 37.5\%$. Use terms that a person who has not studied statistics could understand.

h. (1) Compute the 95% confidence interval for the mean percentage of body fat for men who weigh 180 lbs.

(2) Explain the interpretation of this confidence interval in this context. Use terms that are understandable to the general public.

i. (1) Compute the 95% prediction interval for the percentage of body fat of men who weigh 180 lbs.

(2) Explain the interpretation of this prediction interval in this context. Use terms that are understandable to the general public.

j. Why is the 95% prediction interval wider than the 95% confidence interval? Use terms that are understandable to the general public.

k. Describe how the margins of error for the 95% confidence and prediction intervals for the percentage of body fat of men who weigh 90 pounds would differ from those computed for men who weigh 180 pounds. (You don't have to compute these intervals to answer the question.) Would you have the same confidence in the accuracy of the intervals computed for the 90-pound man as you do for the intervals for the 180-pound man? Explain.

l. Compute the correlation coefficient for the association between percentage of body fat and weight *based on information in the regression print-out*. Perform a test of significance for this correlation coefficient r.

(1) State H_0 and H_a.

(2) Describe the sampling distribution of r (center, spread, shape) under H_0.

(3) Compute the test statistic and determine the p-value.

(4) Explain the p-value in the context of this study. Use terms understandable to the general public.

(5) State your conclusion with regard to the original question, in a format suitable for a scientific paper.

(6) Compare the outcome of this test of significance with the outcome of the regression analysis. Should these outcomes be similar?

m. Do the data used in this analysis meet the assumptions for regression and correlation analysis? State each assumption, indicate if the data meet the assumption, and explicitly refer to those parts of the print-out that you used to make your determinations.

Chapter 13 Study Problem

1. Coral reefs in the Caribbean Sea have been in a severe state of decline in recent decades. Although there are many factors involved, an important cause of this decline appears to be widespread mortality of one species of sea urchin (a spiny ball-like organism). These sea urchins graze on algae that cover coral reefs and prevent coral larvae from attaching to the reef. When the sea urchin population crashed, the percentage of the reef covered by algae increased. As part of a recovery effort, researchers wanted to better define the association between the abundance of this sea urchin species and the percentage of the reef covered by algae. They obtained field data from 24 coral reef locations around Jamaica. At each location they randomly located a 5 m × 5 m square plot. In each of these plots they recorded the number of sea urchins present and the percentage of the coral reef that was covered by algae. The results of this study are presented below. The investigators will use these data to define a regression equation that can predict the percentage of algae cover from sea urchin abundance (number per m²).

Data Analysis

Variable	N	Mean	Median	SD	Min	Max
Algae	24	21.1	22.5	10.9	2.5	41.4
NUrchins	24	6.37	5.9	3.61	0.4	12.9

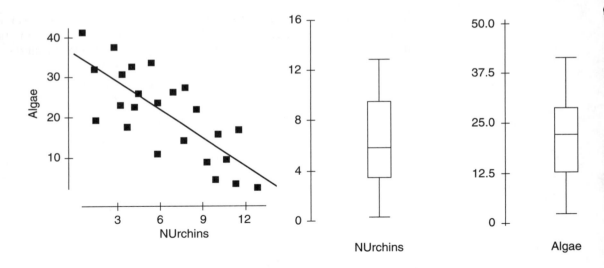

Dependent variable is: Algae
R-squared $= 59.5\%$ R-squared (adjusted) $= 57.7\%$
$s = 7.078$ with $24 - 2 = 22$ degrees of freedom

Source	Sum of Squares	df	Mean Square	F-ratio
Regression	1622.44	1	1622.44	32.4
Residual	1102.22	22	50.1	

Variable	Coefficient	s.e. of Coeff	t-ratio	Prob
Constant	35.9126	2.978	12.1	≤ 0.0001
NUrchins	−2.3279	0.4091	−5.69	≤ 0.0001

Analysis of residuals:

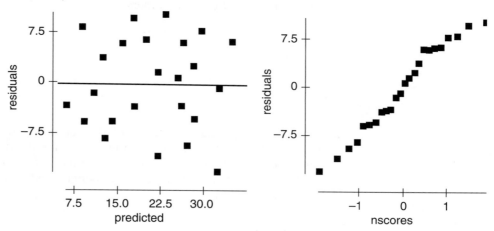

a. Is there an association between the abundance of this sea urchin species and the percentage of coral reef surface area covered by algae? Explain how you arrived at your conclusion with specific reference to parts of the print-out.

b. What do you conclude from the t-test for the Y-intercept in the coefficients table at the bottom of the regression print-out? How do you interpret this result in biological terms?

c. What do you conclude from the *t*-test for the slope coefficient in the coefficients table? How do you interpret this result in biological terms?

d. What can you conclude from the ANOVA *F*-test?

e. Explain the meaning of R^2 − adjusted = 57.7%. Use terms that a person who has not studied statistics could understand.

f. Compute the predicted percentage of algae cover and the residual for the following data values selected from the data used to compute the LSR line. *Note:* Although there is no such thing as a fractional sea urchin, abundance is expressed as number per m^2, computed as the total number counted in the 5 m × 5 m plot, divided by plot area of 25 m^2.

Number of Sea Urchins (*X*)	0.4	5.9	10.7
Percentage of Cover of Algae (*Y*)	41.4	11.2	9.6
Predicted Algae Cover ($\hat{y}$)	_____	_____	_____
Residual	_____	_____	_____

g. Compute 95% confidence interval for *X* = 10.7 sea urchins per m^2.

h. Explain the interpretation of this confidence interval in this context. Use terms that are understandable to the general public.

i. Compute the 95% *prediction* interval for the percentage cover of algae when there are 10.7 sea urchins per m^2.

j. Explain the interpretation of this prediction interval in this context. Use terms that are understandable to the general public.

k. Why is the 95% prediction interval wider than the 95% confidence interval? Explain in terms that are understandable to the general public.

l. Describe how the margins of error for the 95% confidence and prediction intervals for the percentage of algae cover with 10.7 sea urchins per m^2 would differ from the margins of error for predicted algae cover with 5.9 sea urchins per m^2. Explain the reason for this difference. *Note:* You do *not* have to compute the intervals for $X = 5.9$ sea urchins. You should be able to answer this question based solely on your understanding of confidence and prediction intervals and how they should differ for $X = 5.9$ vs. $X = 10.7$.

m. Would you have the same confidence in the accuracy of confidence and prediction intervals computed for $X = 5.9$ sea urchins as for $X = 10.7$ sea urchins? Explain.

n. Would you have the same confidence in the accuracy of confidence and prediction intervals computed for $X = 20$ sea urchins as for $X = 10.7$ sea urchins? Explain.

o. Compute the correlation coefficient for the association between sea urchin abundance and the percentage of coral reef surface area covered by algae *based on information in the regression print-out*. Perform a test of significance for the correlation coefficient r.

r = _____

(1) State H_0 and H_a.

(2) Describe the sampling distribution of r (center, spread, shape) under H_0.

(3) Compute the test statistic and determine the p-value.

(4) Explain the interpretation of this p-value in the context of this study. Use terms understandable to the general public.

Name _____   Date _____

(5) State your conclusion with regard to the original question. Use a format appropriate for a scientific paper.

(6) Compare the outcome of this test of significance with the outcome of the regression analysis. Should these outcomes be similar? Explain.

p. Do the data used in this analysis meet the assumptions for regression and correlation analysis? State each assumption, indicate if the data meet the assumption, and explicitly refer to those parts of the print-out that you used to make your determinations.

Answer Key for Chapter 13 Study Problems

1. **a.** There is a negative, linear association between the abundance of this sea urchin species and the percentage of coral reef surface area covered by algae. This association is clearly displayed in the scatterplot. The slope coefficient in the regression print-out is significantly less than zero, indicating a negative association.

   b. The Y-intercept is significantly greater than zero. When there are zero sea urchins present, the average percentage of algae cover is predicted to be approximately 36%.

   c. The slope coefficient is significantly less than zero. This indicates there is a negative association between sea urchin abundance and the percentage of algae cover on coral reefs. Specifically, for each additional sea urchin per m^2, the percentage of cover of algae decreases by 2.33.

   d. The ANOVA F-test indicates there is sufficient evidence to support the claim for an association between sea urchin abundance and percentage of algae cover on coral reefs.

   e. R^2 adjusted = 57.7% means that 57.7% of the observed variation in the percentage of cover of algae on coral reefs is explained by the association with sea urchin abundance.

   f. Algae (%) = $35.9 - 2.33$ (NUrchins)

Number of Sea Urchins (X)	0.4	5.9	10.7
Percentage of Cover of Algae (Y)	41.4	11.2	9.6
Predicted Algae Cover ($\hat{y}$)	35.0	22.2	11.0
Residual	6.4	−11.0	−1.4

   g. 95% confidence interval for $x = 10.7$ sea urchins per m^2:

$$\hat{y}_i \pm t_{0.025, \text{df} = 22} \; s \sqrt{\frac{1}{n} + \frac{(x_i - \bar{x})^2}{(n-1)S_x^2}}$$

$$= 11.0 \pm 2.074 \, (7.078) \sqrt{\frac{1}{24} + \frac{(10.7 - 6.37)^2}{(24-1)(3.61)^2}}$$

$$= 11.0 \pm 2.074 \, (7.078) \sqrt{0.0417 + 0.0626}$$

$$= \mathbf{11.0 \pm 4.74}$$

   h. *Interpretation:* Because 95% of all intervals computed by this method include the true *population mean*, we are 95% confident that the true mean of the percentage of algae cover on coral reefs where there are an average of 10.7 sea urchins per m^2 is within the range of 11 ± 4.74.

i. 95% prediction interval for $x = 10.7$ sea urchins per m^2:

$$\hat{y}_i \pm t_{0.025, df=22} \; s \sqrt{1 + \frac{1}{n} + \frac{(x_i - \bar{x})^2}{(n-1)S_x^2}}$$

$$= 11.0 \pm 2.074 \; (7.078) \sqrt{1 + \frac{1}{24} + \frac{(10.7 - 6.37)^2}{(24 - 1)(3.61)^2}}$$

$$= 11.0 \pm 2.074 \; (7.078) \sqrt{1 + 0.0417 + 0.0626}$$

$$= \mathbf{11.0 \pm 15.4}$$

j. *Interpretation:* Because 95% of all intervals computed by this method include the true value for an *individual*, we are 95% confident that the true percentage of algae cover on a specific coral reef where there is an average of 10.7 sea urchins per m^2 is within the range of 11 ± 15.4.

k. The 95% prediction interval is wider than the 95% confidence interval because the confidence interval is for the *mean* algae abundance for all reefs with $x = 10.7$ sea urchins per m^2, whereas the prediction interval is for a specific individual reef with the same sea urchin abundance. Individuals are always more variable than the mean of a sample of individuals. Hence, a wider interval is required to maintain the same confidence that the true value is contained therein.

l. The margins of error for the 95% confidence and prediction intervals for the percentage of algae cover with 10.7 sea urchins per m^2 would be wider than the corresponding intervals for predicted algae cover with 5.9 sea urchins per m^2. The value $x = 5.9$ is close to the mean for the sea urchin abundance data, whereas $x = 10.7$ is out near the upper end of the range of the data. Random sampling variation in the value of the slope of the LSR line causes there to be more random variation (less precision) for estimates near the ends of the LSR line than for estimates for x-values near the middle of the line.

m. We can place the same level of confidence that the confidence intervals will include the true values for algae cover given sea urchin abundance of $x = 5.9$ and $x = 10.7$. Both of these x-values are within the range of the data used to compute the LSR line. The only difference is that the estimate for $x = 5.9$ is more precise, as described in part (l).

n. We *cannot* have the same confidence in the accuracy of confidence and prediction intervals computed for $x = 20$ sea urchins as for $x = 10.7$ sea urchins. The value $x = 20$ is beyond the range of the data used to compute the LSR line. We have no way of knowing if this association remains linear at higher abundance of sea urchins. If the true association became nonlinear at x-values beyond the range of the data, estimates of algae cover based on the LSR line could be biased.

o. Correlation coefficient for the association between sea urchin abundance and the percentage of coral reefs covered by algae:

$$\sqrt{R^2} = \sqrt{0.595} = -0.77 \text{ } Note: \text{ Sign is based on sign of slope of LSR line.}$$

(1) $H_0: \rho = 0$

   $H_a: \rho \neq 0$

(2) $E(r) = 0$

   $S_r = 0.136 = \sqrt{(1 - 0.77^2)/(24 - 2)}$

   Shape is Normal (since sea urchin abundance and algae cover are both Normal).

(3) $t_{test} = -0.77 / 0.136 = -5.66$

   $p < 0.001 \text{ } (= 0.0005 \times 2)$

(4) The probability of obtaining the observed correlation (-0.77) between sea urchin abundance and the percentage cover of algae on coral reefs due only to random variation is less than 1 time out of 1,000. Since this is so unlikely to be due just to random variation, this result provides strong evidence that there is a negative association between sea urchin abundance and algae cover.

(5) *Conclusion:* There is a negative association between sea urchin abundance and percent cover of algae on coral reefs on the coast of Jamaica ($r = -0.77$, $p < 0.001$).

(6) The t-test values for Pearson's correlation coefficient r and the slope of the regression line b_1 are virtually identical. These two statistics are alternate ways of assessing the same thing (i.e., the strength and nature of the association).

p. Evaluation of assumptions:

Randomized, unbiased study design: Plots were randomly located on reefs. We can only assume that measurements of sea urchin abundance and the algae percentage of cover were performed using appropriate methods.

True association between sea urchin abundance and algae cover is linear: The scatterplot and residuals vs. predicted plot both indicate that the association is linear.

The variance of algae cover is constant across the range of sea urchin abundance: There is no indication in the residuals vs. predicted plot of a consistent pattern of increasing or decreasing variation of the data points about the regression as the value of X increases. This assumption is valid.

The residuals are Normally distributed: The points in the Normal quantile plot for the residuals fall more or less along a straight line. This assumption is valid.

Exercise A: Regression Analysis

Objective

1. You will determine if a data set meets the assumptions for using linear regression analysis to describe the association between two quantitative variables.

2. You will perform linear regression analysis, interpret the results, and compute confidence and prediction intervals for estimates from the regression model.

Introduction

The primary use of regression analysis of the association between two variables is to derive an equation that can be used to predict values for one variable based on knowledge of the values of the other variable. In some cases, one variable is difficult to measure and the other variable is more easily measured. To save effort that would otherwise be required to measure the "difficult" variable, a study is done to quantify the association between the difficult and "easy" variables. If the association is reasonably strong, data are obtained for the easily measured variable, and a regression equation is used to estimate values for the difficult variable from the measurements of the easy variable.

A common practice in forestry science is to use regression to describe the association between tree stem diameter at 1.4 m aboveground (called "breast height") and other characteristics of the tree that are more difficult to measure (age, height, wood volume). Diameter can easily be measured in seconds, but determining tree age, height, or volume requires much more time and effort. For example, tree age can be determined only by obtaining core samples from the tree trunk and counting the annual growth rings (a very laborious task). In many research projects, data for hundreds or even thousands of trees are required. If these more difficult variables had to be measured directly on each sample tree, the amount of field time and effort would be significantly increased, at a substantial cost. Fortunately, there is often a strong association between the diameter of the stem at breast height (DBH) and tree age, height, and volume. Thus, if we develop a regression equation that estimates values for age, height, and volume from a DBH measurement, we could more efficiently gather information for these variables. In practice, regression equations have been developed for many commercially important tree species. (The association between DBH and age, height, and volume often differs between trees species with different morphology.) Often, the trees used to develop these equations are harvested to enable the most accurate measurements of tree age, height, and volume. Regression equations developed from these small samples of trees are then used to predict age, height, or volume from DBH for thousands of trees measured during forest inventory.

During this exercise you will: (1) Use scatterplots to explore the nature of the association between tree age and tree DBH; (2) perform linear regression analysis to quantify the age-DBH association, and interpret the resulting computer print-outs; (3) perform graphical tests to determine if data meet the assumptions of regression analysis; and (4) compute predicted age, with associated confidence and prediction intervals, when given a value of DBH.

Instructions

1. Obtain the **DBH-Age** data file.

2. Compute the summary statistics for DBH and age (at least sample size, mean, standard deviation, minimum, and maximum).

3. Make a scatterplot of Age (*Y*) and DBH (*X*). If your statistics software provides it, add a linear regression line to the scatterplot to help you see any trend.

4. Perform linear regression analyses for Age (Y) on DBH (X).

5. Create graphs of residuals to assess whether or not these data meet the assumptions for least squares regression, including:

    a. Plot of residuals vs. predicted (or residuals vs. the X-variable DBH).

    b. Normal quantile plot of the residuals.

Report   Put the summary statistics for DBH and age, the scatterplot of age on DBH, the regression print-out, the residuals vs. predicted scatterplot, and the residuals Normal quantile plot on one or two word processor pages, with your name and a title at the top.

Type answers to the following questions in *complete sentences and paragraphs*.

1. Do the data meet the assumptions for linear regression analysis? Describe how you evaluated the print-outs to make these determinations, including reference to all the relevant statistics or graphs that form the basis of your determination.

2. Describe the interpretation of the F-value in the Analysis of Variance table in the context of this specific study. Use terms understandable to the general public. What conclusion could you draw based on the results presented in the regression ANOVA table?

3. Below the ANOVA table of the regression print-out is a table presenting the Y-intercept (sometimes called *constant*) and the slope coefficient for the regression equation. Each of these estimates has a standard error, a t-value, and a p-value. *For each of these two t-tests*, explain the meaning of the t-test p-value in the context of this study. Use terms understandable to the general public and state the appropriate conclusion.

4. For a tree with DBH = 40 cm, compute the estimated age using the regression equation *and* compute the 95% confidence interval and 95% prediction interval for this estimate. Show all calculations. Explain the interpretation of the confidence and prediction intervals in the context of this study. Use terms understandable to the general public. *Note:* Calculations may be neatly handwritten.

5. Describe the difference between a confidence interval and a prediction interval.

6. Use the regression equation to compute estimated age for a tree with DBH = 60 cm.

7. Are you more, less, or equally confident in the age estimates for trees with 40 cm and 60 cm DBH? Explain your answer.

8. Describe the interpretation of the R^2 value presented in the regression print-out in the context of this study. Use terms understandable to the general public.

9. Using the regression print-out, compute the Pearson's correlation coefficient for the associations between DBH and age. Describe how you determined the sign of the correlation.

i. For each cell in the contingency table, compute the adjusted residual and enter this value in the third blank line of each cell. Show your calculations below.

j. State your conclusion with regard to the original scientific question. Was there a difference in the ultimate survival rate of patients treated with the different CPR methods? If so, describe the nature of this association based on the adjusted residuals.

k. Compute the 95% confidence intervals for the proportions of cardiac arrest patients that survive after chest compression only CPR and full CPR.

Full CPR: **Chest Compression Only:**

3. Cigarette smoking has been shown to be a risk factor for complications during and after surgery. Mechanisms by which smoking might cause these complications include effects on pulmonary and/or cardiac function, reduced immune function, and reduced collagen production. Medical researchers wanted to determine if patients who reduce or stop smoking before surgery can reduce their risk of complications. The researchers identified 120 patients who smoked daily that were scheduled for hip or knee replacement surgery. Equal numbers of these patients were randomly assigned to two groups: Control (nothing done about their smoking) and Treatment (received smoking cessation treatment, including nicotine substitution patches). All individuals in the treatment group either stopped smoking entirely or reduced their smoking by at least 50% for a 6- to 8-week period leading up to their joint replacement surgery. An assessor who did not know the group assignments evaluated each patient for wound-related complications within four weeks after surgery. Of the 60 patients assigned to each experimental group, some either postponed or canceled their surgery; $n = 52$ in the control group completed the experiment, and $n = 56$ in the treatment group. (Based on a study by Møller and others, 2002. *The Lancet* 359: 114–117.)

Obtain the data file **Ch14Pr3**. Use a statistics computer program to perform a χ^2 test of independence to evaluate if the results of this study provide evidence that smoking influences the occurrence of wound complications after surgery. The contingency table should meet the following format specification:

- Smoking Cessation (Yes, No) should be the column variable and Wound Complication (Yes, No) should be the row variable.

- You should specify that each cell in the table contain the following statistics: count, column percentage (or proportion), expected value, standardized residual, marginal counts, and proportions for both variables.

- If necessary, you should specify that you want the χ^2 test and associated *p*-value.

- Copy/Paste the statistical print-out to a word processor and type answers to the following questions below the print-out.

    a. State the H_0 and H_a appropriate for this study.

    b. Interpret the meaning of the *p*-value from the χ^2 test in the context of this scientific question. Use terms understandable to the general public.

    c. State your conclusion with regard to the original scientific question. Use the adjusted residuals to describe the nature of the association.

Chapter 14 Supplemental Problem

1. There is substantial evidence that long-term survival of African-Americans after kidney transplantation is lower than for Caucasian Americans. Possible reasons for this difference include inferior histocompatibility matching, poor compliance with post-operative treatment regimes, and lower socioeconomic status. Medical researchers wanted to determine if differences in survival after organ transplantation extended to other racial or ethnic groups and types of organ transplants. They obtain data from the United Network of Organ Sharing (UNOS) for all liver transplants done between 1988 and 1996. Complete information on donor, recipient, and post-operative outcomes was available for 16,669 individuals. The variable of interest was survival rate of patients

after two years. The results are presented below. (Based on a study by Nair and others, 2002. *The Lancet* 359: 287–292.)

Alive after 2 yrs ↓	Caucasian	African-American	Hispanic	Asian-American	Row Totals
Race/Ethnic Group					
Yes	11287	771	1212	341	13611
Exp	_____	_____	_____	_____	$\hat{p}_r$ = _____
$(O-E)^2/E$	_____	_____	_____	_____	
Adj. res	_____	_____	_____	_____	
No	2312	271	322	153	3058
Exp	_____	_____	_____	_____	$\hat{p}_r$ = _____
$(O-E)^2/E$	_____	_____	_____	_____	
Adj. res	_____	_____	_____	_____	
Column Totals	13599 $\hat{p}_c$ = _____	1042 $\hat{p}_c$ = _____	1534 $\hat{p}_c$ = _____	494 $\hat{p}_c$ = _____	16669

a. State the H_0 and H_a appropriate for this study, using appropriate subscripts to clearly identify the populations.

b. Assess whether or not this study and results fulfill assumptions for a χ^2 test of independence.

c. Compute the row and column proportions and enter these values in the spaces provided in the contingency table above.

d. Using the row and column proportions, compute the expected counts of individuals for each of the cells in the contingency table, assuming H_0 is true. Enter these values in the first blank line in each cell.

e. For each cell in the contingency table, compute the quantity $(O - E)^2/E$. Enter these values in the second blank line in each cell of the table above.

f. Sum the values for $(O - E)^2/E$ across the cells in the contingency table to compute the χ^2 test statistic. Enter this value here: χ^2 = _____

g. State the number of degrees of freedom for this analysis. df = _____

h. Use Appendix Table 8 to determine the approximate *p*-value associated with the computed χ^2 statistic. Enter this *p*-value here: $p < $ _____

i. Use a computer spreadsheet program to determine the exact *p*-value associated with the computed χ^2 statistic (see Ch. 12). Enter this *p*-value here: $p = $ _____

Name _____   Date _____

j. For each cell in the contingency table, compute the adjusted residual and enter this value in the third blank line of each cell. Show your calculations below.

k. State your conclusion with regard to the original scientific question. Is long-term survival after liver transplantation influenced by the ethnic background of the patient? Describe the nature of this association based on the adjusted residuals.

l. Compute the 95% confidence intervals for the proportions of patients in each group who survive at least two years after liver transplantation.

Caucasian:

African-American:

Hispanic:

Asian-American:

m. Explain what these confidence intervals mean in this context. Use terms under-standable to someone who has not studied statistics.

Chapter 14 Study Problem

1. Sickle-cell anemia is a debilitating genetic disease found in African populations. Individuals that have the defective gene on both chromosomes (homozygous recessive) typically die at an early age. Nonetheless, the prevalence of the defective gene remains high in African populations. One explanation that has been proposed for the continued high prevalence of the sickle-cell anemia gene is that individuals who have one normal and one defective gene (heterozygous) are somehow more resistant to the malaria parasite than individuals who do not carry the defective gene. Malaria is also a debilitating disease that causes early mortality, killing more people worldwide than any other disease. Heterozygous individuals can pass on the sickle-cell disease to their children but are not themselves debilitated by the sickle-cell gene because their one normal copy of the gene performs the necessary function. The following data were obtained from a study of the association between the presence of the sickle-cell gene and the severity of malarial infection. I use **N** to denote a normal gene and **S** to denote the sickle-cell gene. Every person has two copies of this gene, one on each of a pair of chromosomes. N/S denotes the heterozygous genotype (carrier), N/N the homozygous normal condition. Perform a test of significance to determine if these data provide evidence that resistance to malaria is related to the presence of the sickle-cell anemia gene.

Malarial Infection			
Sickle-cell Genotype	Severe	None/Light	Row Totals
N/S	36	100	136
Exp	_____	_____	$\hat{p}_r =$ _____
$(O-E)^2/E$	_____	_____	
Adj. res	_____	_____	
N/N (Non-carrier Normal)	152	255	407
Exp	_____	_____	$\hat{p}_r =$ _____
$(O-E)^2/E$	_____	_____	
Adj. res	_____	_____	
Column Totals	188 $\hat{p}_c =$ _____	355 $\hat{p}_c =$ _____	543

a. State the H_0 and H_a appropriate for this study, using appropriate subscripts to clearly identify the populations.

b. Compute the row and column proportions and enter these values in the spaces provided in the contingency table above.

c. Using the row and column proportions, compute the expected counts of individuals for each of the cells in the contingency table, assuming H_0 is true. Enter these values in the first blank line in each cell.

d. For each cell in the contingency table, compute the quantity $(O - E)^2/E$. Enter these values in the second blank line in each cell of the table above.

e. Sum the values for $(O - E)^2/E$ across the cells in the contingency table to compute the χ^2 test statistic. Enter this value here: $\chi^2 =$ _____

f. State the number of degrees of freedom for this analysis. df = _____

g. Use Appendix Table 8 to determine the approximate p-value associated with the computed χ^2 statistic. Enter this p-value here: $p <$ _____

h. Use a computer spreadsheet program to determine the exact p-value associated with the computed χ^2 statistic (see Ch. 12). Enter this p-value here: $p =$ _____

i. For each cell in the contingency table, compute the adjusted residual and enter this value in the third blank line of each cell. Show your calculations below.

j. State your conclusion with regard to the original scientific question. If there is sufficient evidence to infer there is an association, use the adjusted residuals to describe the nature of the association. That is, are carriers of the sickle-cell gene less likely to suffer from severe malarial infection?

Answer Key for Chapter 14 Study Problem

1. a. $H_0: P_{Sc\times M} = P_{Sc} \times P_M$ (Events independent)
$H_a: P_{Sc\times M} \neq P_{Sc} \times P_M$ (Events *not* independent)

b. to d.

Sickle-cell Genotype	Malarial Infection		Row Totals
	Severe	**None/Light**	
N/S (Sickle-cell Carrier)	36	100	136
Exp	**47**	**89**	**0.2505**
$(O-E)^2/E$	**2.57**	**1.36**	
Adj. res	**−2.29**	**2.29**	
N/N (Non-carrier Normal)	152	255	407
Exp	**141**	**266**	**0.7495**
$(O-E)^2/E$	**0.86**	**0.45**	
Adj. res	**2.29**	**−2.29**	
Column Totals	188 **0.3462**	355 **0.6538**	543

e. $\chi^2 = (36 - 47)^2/47 + (100-89)^2/89 + (152 - 141)^2/141 + (255 - 266)^2/266$
$= 2.57 + 1.36 + 0.86 + 0.45 = \mathbf{5.24}$

f. df = (#rows − 1)(#cols − 1) = (2 − 1)(2 − 1) = 1

g. $p < \mathbf{0.025}$

h. $p = \mathbf{0.0221}$

i. Adjusted Residuals

N/S × Severe $(36 - 47) / \sqrt{47(1 - 0.2505)(1 - 0.3462)} = -11 / 4.80 = \mathbf{-2.29}$

N/S × Light $(100 - 89) / \sqrt{89(1 - 0.2505)(1 - 0.6538)} = 11 / 4.80 = \mathbf{2.29}$

N/N × Severe $(152 - 141) / \sqrt{141(1 - 0.7495)(1 - 0.3462)} = 11 / 4.80 = \mathbf{2.29}$

N/N × Light $(255 - 266) / \sqrt{266(1 - 0.7495)(1 - 0.6538)} = -11 / 4.80 = \mathbf{-2.29}$

j. *Conclusion:* Severity of malarial infection depended on whether or not an individual carried the sickle-cell gene (p = 0.0221). Carriers of the sickle-cell gene were less likely to suffer from severe malarial infection, as indicated by the negative residuals. Individuals who didn't have the sickle-cell gene were more likely to suffer from severe malarial infection, as indicated by the positive standardized residual.

Exercise A: Contingency Table Analysis

Objective
1. You will perform contingency table analysis to test hypotheses regarding the independence of two categorical variables.
2. You will interpret contingency tables when the overall Null hypothesis of independence is rejected, to identify the nature of the association.

Introduction
It is common in biology that we wish to explore associations between variables that are categorical. Examples: Are adult male white-tail deer more likely to die than adult female deer during the winter after the rut season? Are brightly colored (e.g., yellow, red) flowers more likely to be visited by pollinators than green-colored flowers? In some cases, the categorical variable may be quantitative or at least have a logical ordering of classes (e.g., age classes). However, in many cases there is no logical ordering of the classes or categories (e.g., flower color, gender).

The analysis of associations among categorical variables is generally accomplished using a procedure called contingency table analysis. Contingency table analysis is used to determine if the distribution of individuals in various classes of one categorical variable is independent of their characteristics with regard to another categorical variable. The mechanics of contingency table analysis are similar to those described for two-way table analysis (Chapter 12), but the emphasis is on whether or not there is an *association* between two or more categorical variables, rather than a *difference* among two or more populations. Sometimes the distinction between contingency and two-way table analyses is subtle. The only difference is in the nature of the Null and Alternative hypotheses and the statement of your conclusion from the test.

In this exercise you will perform contingency table analysis for the association between local soil and topographic conditions and the response of oak trees to severe drought. Severe droughts often injure trees in ways that reduce their capacity for growth over some period of time, even many years after the actual drought is over. During this period of reduced growth after droughts, trees may be more susceptible to attack by a number of insect pests that are capable of killing only weakened trees. The response variable in this research is **Decline Duration**, defined as the number of consecutive years during and after a severe drought in 1953–54 when tree growth was below the average growth for the five-year period prior to the drought. The larger the value of this variable, the longer the tree suffered reduced growth and increased susceptibility to disease and death. Hence, individuals with large values for this variable were more negatively affected by the drought than individuals with small values (which indicate rapid recovery).

Drought-site interaction study
White and black oak trees were randomly sampled across a range of site conditions defined by soil pH and slope. Low soil pH is generally associated with low nutrient availability. South-facing slopes near the top of a hill are dry; North-facing slopes near the bottom of a hill are moist. Sampling was done on sites that represent four combinations of slope position (moisture) and soil pH (nutrients), as described in Table 14.1.

TABLE 14.1   Site types for the Hoosier National Forest Drought–Site Interaction
Study

Site ID Code	Site Characteristics
LpHLN	Site with **low soil pH** on **low-north** facing slopes. (Higher soil water availability + lower temperature = lower drought stress.)
LpHHS	Site with **low soil pH** on **high-south** facing slopes. (Lower soil water availability + higher temperature = higher drought stress.)
HpHLN	Site with **higher soil pH** on **low-north** facing slopes. (Higher soil water availability + lower temperature = lower drought stress.)
HpHHS	Site with **higher soil pH** on **high-south** facing slopes. (Lower soil water availability + higher temperature = higher drought stress.)

These four combinations of two levels for each of two site factors represent a *factorial design* to study the interaction of soil pH and slope-aspect effects on oak growth responses to drought. If more trees are adversely affected on LpH sites than HpH sites, this would indicate that low soil pH (low nutrient availability) makes trees more sensitive to droughts. If more trees are adversely affected on HS sites than on LN sites, this would indicate that trees on hot-dry high-south sites are more sensitive to droughts than trees on cooler, moister low-north sites.

In this exercise you will use contingency table analysis to address questions about factors that influence post-drought decline duration through analysis of contingency tables (χ^2 test and standardized residuals) and through subset χ^2 analyses. The specific question addressed by these analyses is: Did the local site environment (soil pH, slope-aspect) influence the post-drought growth decline duration for these oak species?

Instructions

1. The data file **Ch14Exercise** contains data for the following variables, recorded for each individual oak tree in the sample: Site ID, Soil pH class, Slope-Aspect class, Species code (WO = white oak, BO = black oak) and Decline Duration.

2. The variable **Decline** contains the exact number of consecutive years after the drought that the individual trees exhibited below average growth. You must convert the more or less continuous, quantitative **Decline** into a numeric categorical variable **Dclass** with a relatively small number of discrete classes. See the instructions in the Chapter 12 exercise.

3. Before performing the contingency table analysis, select the kinds of information you want in your table. You should specify the following: count (observed), expected values, standardized residuals, χ^2 value, row margins, and column margins.

4. The general procedure for performing contingency table analysis requires that you select the two variables (**Site** and **Dclass** in this analysis) and then select **Contingency Table** analysis from a menu of statistical tests. This item may be in a submenu titled **Tables**, and may also be called **Cross Tabulation**. Each statistics computer program has different menu structures. You should consult the Chapter 14 Tutorial for your statistics software.

5. If the overall χ^2 test for site vs. Dclass is significant, you must first determine if tree decline is influenced by soil pH, slope position or both. You can perform a subset χ^2 test using the variables **Soil pH** and **Slope**.

Name _____   Date _____

    a. Select **Soil pH** and **Dclass**, then request a contingency table. If this test is significant, you can conclude that soil pH contributes to variation in post-drought decline duration. In this case, you should look at the standardized residuals to determine how soil pH affects decline duration. Large + residuals indicate more trees than expected were found in that cell of the table, and vice versa for − residuals.

    b. Select **Slope** and **DClass**, then request a contingency table. If this χ^2 test is significant, you can determine that site slope position contributes to variation in post-drought decline duration. Again, in this case you should look at residuals to explain the nature of this effect.

Report You should have three contingency tables (Dclass vs. Site, pH, and Slope, respectively). Put each table on a word processor page, with your name and an appropriate title for the analysis at the top of the page. On each of these pages, type the Null and Alternative hypotheses, using appropriate subscripts to specify categories and variables. State your conclusion with regard to the scientific question.

    Type answers to the following questions in *complete sentences and paragraphs*. Explain your answers with explicit references to the relevant contingency tables.

    1. Do both slope and soil pH influence post-drought decline duration of oaks?
    2. Based on comparisons of standardized residuals, on which site type is post-drought decline duration greatest (i.e., on which site type do a greater than expected number of trees exhibit prolonged growth decline after a severe drought)?

Tests of Association Between Two Categorical Variables   **14–17**